Peter Wißling

Metallic Effect Pigments

Fundamentals and Applications

Translated by John Haim and David Hyatt

Cover: Eckart GmbH & Co. KG

Peter Wißling et al.:
Metallic Effect Pigments
Hannover: Vincentz Network 2006
European Coatings Literature
ISBN 3-87870-171-3
ISBN 978-3-87870-171-2

Please ask for our book catalogue
Vincentz Network, Plathnerstr. 4c, 30175 Hannover, Germany
T (202) 684-6630, F (202) 380-9129
E-mail: books@american-coatings.com, www.american-coatings.com

Layout: Sperling Info Design GmbH, Gehrden

ISBN 3-87870-171-3
ISBN 978-3-87870-171-2

European Coatings Literature

Peter Wißling

Metallic Effect Pigments

Fundamentals and Applications

Translated by John Haim and David Hyatt

Foreword

Silver cars, shiny metallic mobile phones and televisions, gold labels on beer bottles, glittering lipsticks and sparkling fireworks – these are just a few examples of the everyday applications of metallic pigments which are becoming more widespread and more diverse every day.

Metallic pigments are distinguished by their optical effects. Objects coated with metallic effect pigments appear shiny and metallic. From different angles of observation, a pronounced light/dark or flip-flop effect accentuates the shape of the objects and draws attention to curved surfaces. More than that, metallic pigments based on their reflective properties also contribute to UV protection of coatings. A special group of metal pigments, is used for technical applications such as light-weight concrete, pyrotechnics and corrosion protection where, as a rule, the chemical and physical rather than the optical properties are of most importance.

This book gives a comprehensive overview of metallic effect pigments. Starting with the manufacturing process through technical properties to areas of application, the individual chapters offer condensed specialist knowledge based on many years of experience at the Eckart Company. The importance of metallic effect pigments, their wide range of applications and related specifications as well as their development potential are explained.

The contents are aimed at a broad audience. The fundamentals and applications of metallic effect pigments presented here will certainly help to provide basic knowledge to students and laymen with technical interest. In addition, they will also enable qualified personnel to deepen their knowledge on specific topics.

Dr Klaus Greiwe

Eckart GmbH & Co. KG
Velden/Germany, April 2006

Contents

Part I Historical background

At the beginning of the third millennium, silver metallics have achieved major importance: silver became “Millennium Colour”. Today metallic pigments can be found in a growing number of applications, from kitchen appliances through textile and even to food decoration. The trend towards metallic pigments is reaching a new peak. However, the roots could be tracked five thousand years back in Egypt.

Precious metals – the historical roots

There is clear evidence that in Egypt already in 2500 B.C., gold was hammered out into very thin foils. These foils were then used as decorative coating on stucco building ornamentation and other works of art. Such extravagance could only be applied to the noblest edifices, such as the Pharaohs’ palaces and the temples of the ancient Egyptian gods, since pure gold was (and still is) extremely expensive. This most valuable of all metals imparted a sublime character to the objects it decorated.

Starting from the Middle East, the technique of hammering gold out into thin foil spread via Mesopotamia, India and Korea as far as Japan where it was in evidence from about 800 A.D.

Gold beating shops in Germany

Via Greece and Italy this type of precious decoration reached Germany, and a centre of gold beating handcraft was formed in the Franconian towns of Fürth, Nuremberg and Schwabach. Subsequently, many developments in the field of metallic pigments were made in this region. When hammering gold into extremely thin gold-leaf, gold dust was always produced as a side or waste product and consequently it was used as a pigment.

Gold inks and paints

Gold dust pigments were soon mixed into inks and varnishes, enabling the manufacture of gold coloured printing inks and brushable gold paints which were considerably easier to apply compared to gold-leaf. Now it was also possible to gold-paint objects with uneven surfaces and to achieve shining gold prints on parchment. The principal use of such paints besides decorating buildings and furniture was printing documents and decorating ceramics and leather goods.

Brass as a substitute for gold

For a long time gold decorated ceramics and leatherwear were only affordable for the upper class of society until around 1760. Then the significantly cheaper brass was offered as a substitute to pure gold. This alloy was invented by a master craftsman in Fürth called Albert Huber who had realised that the workability of brass is similar to that of gold unlike bronze, the alloy of copper and tin, which is hard and brittle. Although bronze can be formed into solid objects such as bells and statues, it cannot be hammered out into fine pigments. This is possible with the ductile brass alloy made of copper-zinc , also termed gold bronzes (cf. p. 17).

Initially, pure gold colours and gold bronze colours were used in similar amounts since the brass dust resulting from beating with a hammer could be incorporated into printing inks and varnishes as easily as real gold. At the beginning of the 20th century, the term bronze pigments was widely used for these products.

Because of their significantly lower price, bronze pigments were used in an ever increasing range of applications. The demand was particularly for wallpapers and borders, coming from France, England and other countries in Europe and even from the United States.
The increased demand could only be met by the use of efficient manufacturing equipment. Around 1830 in Nuremberg, Georg Benda invented a bronze grinding mill. This first mechanical apparatus for transforming the foils into fine pigments was quickly adopted. Later, bronze pigments were manufactured using steam power. Steam hammers opened the way to industrial scale production in Franconia from about 1860.

Aluminium pigments for silver effects

The introduction of platelet-like aluminium powder in the 19th century enabled silver effects to be achieved. The key factor for the vigorous growth in aluminium pigment consumption was the Kryolith Electrolytic Process. Invented in 1892, it started industrial production of the hitherto extremely expensive aluminium. In Europe, aluminium powder was produced parallel to gold bronze pigments by hammering out aluminium into thin foil. On the contrary, in Japan and in the USA, aluminium powder was a by-product of aluminium smelting and the manufacture of aluminium foil. Around 1910, an American company invented a new milling process: aluminium granules were processed in large ball mills. Large-scale industrial production of aluminium powders had begun. Initially this process was susceptible to explosions as the pigment was manufactured by dry methods. Safe production of aluminium pigments in large amounts became feasible by introduction of white spirit as a grinding medium by Hall. As it was already the case with gold-bronze pigments, from now on platelet-like aluminium pigments were called silver bronzes, a term still in use today.

New areas of application

After World War II, gold and silver bronzes found new areas of application such as in marine paints, plastics and cosmetics. The introduction of metallics, particularly in the automotive industry, created new styling and design possibilities which are considered essential today.

Part II Fundamentals

1 Metallic effect pigments

Metallic flakes are platelet-shaped metallic effect pigments of aluminium, zinc, copper or brass. According to DIN 55944 they are inorganic pigments with a shiny, metallic character. Their stability can be improved by an organic or inorganic coating.

Metallic effect pigments form only a small group in the extensive metallic pigment family. DIN 55943 [1] defines a *metallic pigment* as a pigment which consists of metals or metal alloys. In practice, in coatings technology, the definition is extended to include the many pigments with a metal atom in any chemical form in the molecule. This includes also inorganic colour pigments with one or more metal atoms in their formula. Important representatives are iron oxides, chrome oxide green, and the somewhat ecologically-unfriendly cadmium and bismuth vanadates, and yellow, orange or red chromate pigments. This definition of metal pigments also includes organic colour pigments containing a metal atom, usually as a central atom. Among this group phthalocyanine and ultramarine blue pigments are widely used.

Many fillers and anticorrosive pigments contain one or more metal atoms. They include zinc sulphide, zinc white and zinc phosphate pigments which are state-of-the-art replacements for ecologically-unfriendly anti-corrosive pigments. Iron oxide mica pigments also belong to this group.

Metal pigments are provided in their pure metallic form; one example for this is zinc dust, which is used worldwide as anticorrosive pigment in zinc-rich primers. Although utilised on a much smaller scale, stainless-steel and nickel pigments can be similarly used. They are usually combined with other anti-corrosive pigments. Ferrites have special properties and they are used as ferro-magnetic component for recording media (storage of data).

This list makes no claims to completeness. Further information on individual pigments can be found in the literature [2, 3].

The metallic effect pigments described in the following chapters are mainly designed to produce optically attractive effects (light-dark or colour flop). In addition, some of the pigments fulfil functional requirements (e.g. zinc flakes in corrosion protection systems). Special multi-phase metallic effect pigments, where aluminium flakes function as a carrier which is subsequently coated, are not considered here. The reader is again referred to the literature [3].

References

[1] DIN 55943, October 2001, p. 1 and 13
[2] Ullmanns Enzyklopädie, Weinheim 1979, Vol 18, p. 629ff and p. 545ff
[3] Kittel, H., Pigmente, Füllstoffe und Farbmetrik, Stuttgart 2003, 2nd edition, Vol. 5

2 Manufacture and properties of aluminium, gold bronze and zinc pigments

Manufacturing methods: aluminium pigment manufacturing in the Hall Process (wet grinding)

Origin and extraction

Aluminium is the most abundant metallic element in Earth's crust. It occurs as oxide in numerous minerals. Among these oxides bauxite, a mixture of iron oxide and hydrated aluminium oxide, is the most important. Rich raw material sources are found in France, Hungary, the USA, Italy, Yugoslavia, China and Australia. Iron impurities in bauxite are removed by a wet chemical purification process and the aluminium metal is then obtained by electrolysis of the melt. This aluminium is produced in the form of bars which can be easily transported and stored.

Atomizing and ball milling

In the manufacturing procedure of aluminium granules, the raw product for aluminium pigments – the bars – are molten in an oven at a temperature of at least 700 °C, slightly above the melting point of aluminium. Modern melting furnaces work according the induction principle. The light metal in the liquid phase is atomized through a nozzle under high pressure (Figure II-2.1).

Figure II-2.1: Atomizing

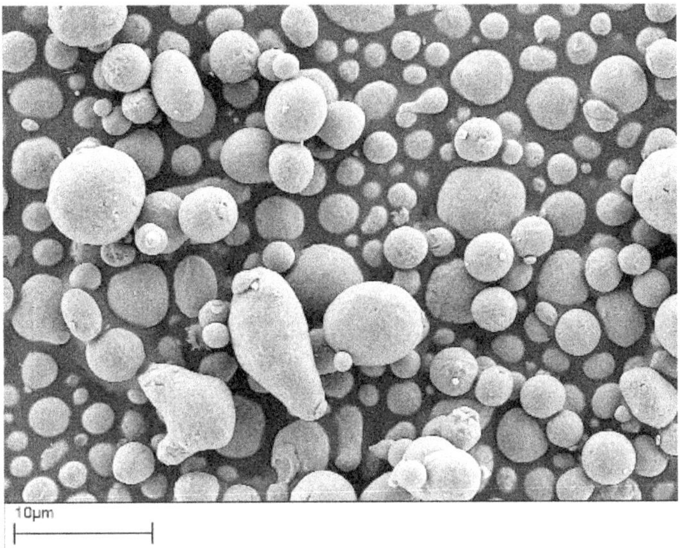

Figure II-2.2: Aluminium Granules

The atomised metal cools down immediately and, depending on the atomizing conditions, is now in the form of isometric granules (Figure II-2.2).

These aluminium granules are the raw material for milling, the process which produces the flake form of metallic effect pigments. Grinding is necessary because the granules are too coarse to be incorporated into thin paint or printing ink films. Furthermore the granules do not have the opacity and gloss required for metallic effect pigments.

In the Hall wet-grinding process, using a ball mill, aluminium granules are mixed with white spirit and a certain amount of lubricant, usually a C_{18}-carboxylic acid such as oleic or stearic acid. The grinding process inside the ball mill (Figure II-2.3) changes the size of the aluminium particles and also converts them into platelets. The lubricant is adsorbed on the highly reactive aluminium surface and coats the hydrophilic oxide layer that is formed immediately after grinding. This avoids uncontrolled reactions such as cold welding of the pigments and formation of agglomerates. The heat produced during the ball milling process is removed by a cooling system.

Sieving

The aluminium pigment consists of flakes with a certain average particle size and particle-size distribution. Agglomerates and coarse particles are removed by sieving (Figure II-2.4).

Figure II-2.3: Ball mill

Figure II-2.4: Sieving process

In addition to the removal of agglomerates which may have been produced in the grinding process, sieving *(German: "Grenzkornsiebung")* has a further important function: it removes oversized particles from the particle size spectrum. In the early days of wet-grinding, target fractions smaller than 63 µm were usual. Later fractions with particle sizes lower than 40 to 45 µm became common. Modern aluminium pigments are sieved using 25 µm mesh. These products have a narrow particle size distribution enabling high degrees of gloss to be achieved in coating applications.

Nowadays these fine-particle metallic effect pigments with narrow particle size distribution are particularly important in the automotive industry where reducing film thicknesses by even 1 µm brings about a marked cost saving. Fine-particle metallic effect pigments are also beneficial in single-coat finishes for mobile phones and computer housings, to name just two applications.

Finishing

For safety reasons, aluminium pigments are sold as pastes. During processing, that is during grinding and subsequent sieving, the concentration of aluminium pigments in the white spirit slurry must not be too high. The aluminium content is usually below 30%. Further on in the pastes however, the amount of aluminium pigment must be as high as possible. The pigment is concentrated in filter presses where excess white spirit is removed. The filter cakes contain 80% aluminium pigment and 20% white spirit.

The commercial product is adjusted to 65% aluminium and 35% solvent using highly compatible aromatic-rich white spirit (e.g. naphtha). These products can be supplied to customers for formulating solvent-based metallic paints or printing inks. Alternatively filter cakes can be further processed to stabilised pigments for use in water-based systems. (cf. Part IV, Chapter 1.1.2). Exchange of solvent for further applications, such as in plastics, is possible too (cf. Part IV, Chapter 3).

The Hametag dry-grinding process for gold bronze pigments

Gold bronze pigments are pigment flakes made from copper or copper-zinc alloys (brass). As the recovery and extraction of zinc are covered in the section on zinc flakes, only copper will be discussed here.

Origin and extraction of copper

Copper is mainly found as oxide, sulphide, chloride or carbonate in the USA, Canada, Chile, the Congo and Zimbabwe. The most important ore is copper pyrites ($CuFeS_2$) which is roasted. In the roasting process in a vertical or reverberate furnace, iron is removed as slag. The remaining copper contains sulphur and is purified by treatment with oxygen. This produces raw copper which is subsequently further purified and refined by electrolysis.

Composition

The starting material for manufacturing gold bronzes is high-purity copper and zinc obtained from an electrolytic process. Both metals are alloyed with addition of some aluminium as a reducing agent. The final colour of the alloy is determined by the copper/zinc ratio. The copper content is generally between 70% and 100%. Because of the alloying, gold bronze pigments have characteristic colours called *natural tones* (Table II-2.1):

Colour name	Alloy composition	Colour
Copper	100% copper	copper red
Pale gold	90% copper, 10% zinc	reddish
Rich pale gold	85% copper, 15% zinc	yellow
Rich gold	70% copper, 30% zinc	greenish yellow

Table II-2.1: Alloy composition for natural tones

Atomizing and grinding

Atomizing the molten metal produces irregular shaped metal granules which are processed in a ball mill with steel balls into gold bronze flakes using stearic acid as an auxiliary. As in the case of aluminium flakes, the lubricant serves primarily to inhibit cold welding of the pigment particles (see above). In contrast to the production of aluminium flakes, the milling of bronze flakes is a dry process known as the Hametag dry-grinding process. This is because dust from gold bronze pigments does not tend to explode in the presence of oxygen and so can be dry milled with suitable safety precautions. The powder produced can be used universally in almost all solvents and binders. Gold bronze pigments are flakes like the aluminium pigments, but their density is about three times higher.

Classification

To obtain different categories of pigment fineness (in some applications gold ink requires coarse particles with average size d_{50} of 50 µm whereas offset printing needs extremely fine particles with a d_{50} less than 5 µm), several grinding steps are necessary under different conditions. Mill size, diameter, rotation speed, ball size, grinding time and lubricant all play important roles. After grinding, cyclones are used to separate different pigment fractions by gravity and centrifugal force. To achieve the best possible metallic gloss from metal pigmented applications, additional lubricant is subsequently "polished" on to the surface of the platelets.

Fire colouring

The varieties of copper and bronze natural tones can be extended by fire colouring before further processing. In this case the pigment is exposed to air for a certain time at a given temperature so that a thin oxide layer is formed on the metal flakes. Interference then produces attractive colour nuances. The most important colours obtainable are lemon, deep gold and fire red.

Zinc flakes

Zinc occurs predominantly in nature as sulphide or oxide ores. Zinc blende, a zinc sulphide, is the most common ore for smelting. It is mined in Poland, France,

England, Australia, Canada and the USA. Zinc blende is first reduced and then purified by electrolysis. As with aluminium flake manufacture, the process involves atomizing the liquid metal using air and then either wet grinding with white spirit or dry milling. Stearic acid is usually used as the lubricant.

Properties

Wetting properties: leafing and non-leafing

In the different manufacturing processes, lubricants are used to prevent cold welding. The chemical composition of the lubricant influences the wetting behaviour of the metal pigment in a characteristic way. With aluminium pigments, a differentiation is made between leafing and non-leafing types (cf. DIN 55923 corresponds ISO 1247). For gold bronze- and zinc pigments, only leafing types are manufactured by the grinding process.

Leafing pigments

Leafing pigments have a high surface tension so that they are not wetted by the binder matrix. Consequently they float in the wet film and orient themselves on top of the surface (Figure II-2.5). This effect is obtained, for example, if stearic acid is used as a milling auxiliary.

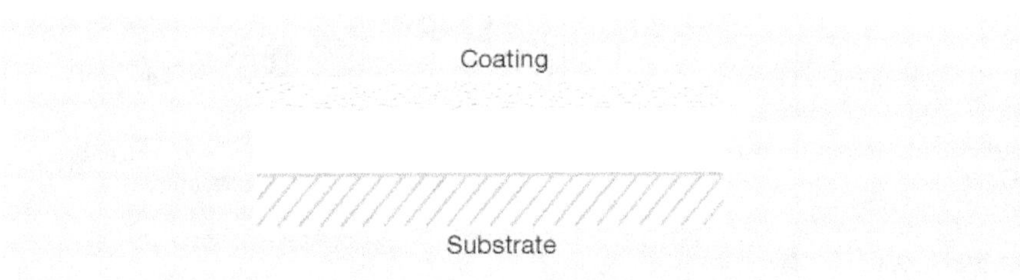

Figure II-2.5: Leafing pigments

Reflection effects

Because of their mirror like effect, leafing pigments are mainly used in decorative paints such as chrome effect coatings and especially in printing inks. While exhibiting the highest gloss, in contrast the rub- and scratch-resistance is just moderate as the flakes are only inadequately fixed in the binder matrix. The most perfect floating behaviour of the leafing pigments is obtained in non-polar systems and aromatic binders. In coatings and inks with strongly polar solvents or binders, there is a tendency for leafing pigments to wet and "drown" and to become non-leafing pigments. If required, especially in print applications, special leafing stabilisers are used.

Barrier protection

Leafing pigments are also ideal anti-corrosive pigments. Because of their floating effect to the surface of a coating they produce a barrier layer which reflects radiation (in the visible, UV and IR light). This barrier layer also hinders penetration of moisture and other reactive media into the paint film or at least, because of the long diffusion pathways, strongly retards it *(tortuosity)*. This protects a paint film and a coated object from over-heating (IR reflection: tank coatings, roof coatings) or strongly retards destruction of the binder by UV rays (cf. p. 148).

Leafing aluminium pigments – an overview

Table II-2.2 contains a general overview of leafing aluminium pigments with important characteristic features and areas of application. (The products used as examples are all manufactured by *Eckart*.)

Fineness	Average particle size in µm	Spreading value cm^2/g	Applications
Coarse	15 to 30	18,000 to 21,000	roof coatings, coatings for containers and tubes (reflecting properties), heat resistant coatings, marine coatings
Medium	8 to 15	21,000 to 40,000	industrial coatings, spray cans
Fine	8 to 5	40,000 to 65,000	chrome effect coatings, decorative coatings, spray cans, reflective coatings (e.g. for vehicles)
Finest	< 6	> 50,000	printing inks
polished (coarse to fine e.g. STANDART® Chromal, STAPA® Luxal	refers to individual product	refers to individual product	highly brilliant chrome effects, reflective coatings, ink applications and powder coatings

Table II-2.2: Leafing pigments

Non-leafing pigments

Non-leafing pigments are those grades which are fully wetted by the binder and therefore distributed uniformly in the whole paint or printing ink film (Fig. II-2.6). The binders are usually polar or highly polar.

The appearance of non-leafing pigments in comparison to leafing grades is not as metallic, but the coating is absolutely rub resistant. Combination with transparent organic pigments (these are coloured pigments which do not scatter light) gives polychromatic metallic paints as used for automotive topcoats (metallics). Furthermore they can be over coated. In some cases addition of strongly polar

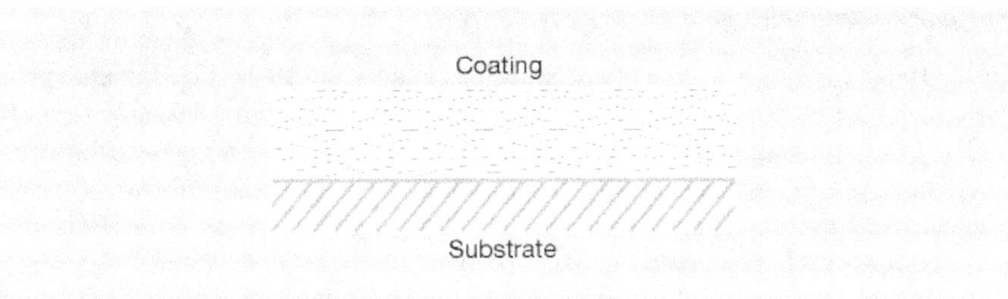

Figure II-2.6: Non-leafing pigments

additives or binders to printing or coating systems can convert leafing pigments into non-leafing pigments. This technique is particularly used in graphic arts to take advantage of the non-leafing behaviour (better abrasion and crack resistance, see above). Customised non-leafing aluminium pigments manufactured using oleic acid as a grinding aid are much more common in today's industry.

Table II-2.3 gives an overview of non-leafing aluminium pigments with important characteristic features and areas of application. (The products used as examples are all manufactured by *Eckart*.)

Pigment type	Particle size distribution	Average particle size in µm	Upper limit of particle size determined by sieving* [µm]	Applications
Cornflakes	very large	10 to 30	< 77 or < 45	coarse to medium types (15 to 30 µm): industrial coatings, cheap automotive coatings, corrosion protection, marine paints, hammer finish paints fine grades (< 15 µm): can/coil coating
Cornflakes e.g. STAPA® Metallic, STAPA® Mobilux, STAPA® Metallux 400, 600, 8000	narrow	10 to 60	< 45 or < 25	automotive coatings (OEM, refinish), plastic coatings, special effect coatings (decorative coatings)
Silver dollars e.g. STAPA® Metallux 2000	very narrow	9 to 35	< 25	automotive and plastic coatings
Shear stable grades e.g. STAPA® NDF	narrow	15 to 35	< 45 or < 25	automotive (OEM)

* sieving with various mesh sizes

Table II-2.3: Non-leafing pigments

Surface coating with gold bronze pigments

Gold bronze pigments, which are produced as leafing grades by the Hametag process, can be given non-leafing properties by post-coating the flakes (e.g. with SiO_2). In addition to conversion of leafing flakes into non-leafing gold bronzes (see above), SiO_2 coating makes the product thermally stable up to 200 °C. Standard leafing grades from the Hametag dry-grinding process are only stable up to about 80 °C. The coated grades are available from *Eckart* under the name STANDART, Resist. Application of a SiO_2 coating also improves the chemical stability of gold bronze pigments as the SiO_2 protective layer prevents de-zincing and inhibits, or at least slows down, dissolution of copper ions. This is especially valuable in aggressive systems such as in binders with high functionality or nitrocellulose systems with high residues of nitric acids. Storage stability is significantly improved by the use of coated non-leafing gold bronzes.

In one-coat paints being in use e.g. for coating plastics (for example for mobile phones or computers) coated non-leafing grades also have advantages as they have significantly improved rub and chemical resistance compared with non-coated grades. The use of the coated grades in stoving enamels has been comparatively unimportant. Here combinations of aluminium flakes with colour pigments offer a wider performance spectrum. The insufficient weathering resistance and tendency to settle in circulation lines in the automotive industry of coated non-leafing gold bronze pigments mean that they have not been processed successfully in the automotive field. At present *Eckart* offers improved SiO_2-coated pigments under the trade name Dorolan® gold bronzes. A relatively dense SiO_2-layer is deposited on the pigment surface by a gentle sol-gel process. This allows stabilisation of fire-coloured bronzes. Further, thanks to the pore-free SiO_2-encapsulation, these new products are also chemically more stable.

Powder coatings are a classic field of application for the mentioned coated gold bronze pigments. The different product groups are discussed in Chapter 1.1.3. in Part IV.

Pigment orientation

For non-leafing metallic effect pigments, orientation of the pigment flake in the medium is of decisive importance for a proper metallic effect finish. The maximum gloss is obtained when the flakes are aligned parallel to the substrate surface. Spot damages from this ideal orientation – especially on large areas – result in a cloudy appearance or the so called *salt- and pepper effect*. In coatings field, the orientation of the flake-shaped pigments is mainly determined by the formulation and the conditions of application. In liquid coatings spray application, when the droplets of the coating are formed more or less quickly, the relatively low spray viscosity results in an initial disorder of the pigments. Immediately after that, fast evaporation of the solvent shrinks the wet film and forces the pigment flake into a position parallel to the surface. The lower the viscosity of the coating film (that is, in general, the higher the proportion of solvent), the more marked is this effect. In turn, this explains why pigment orientation and therefore metallic appearance is much easier achieved in low-solid coatings than in high-solid coatings (cf. p. 81, Figure IV-1.1.1.1 [Part IV, Chapter 1.1.1: Pigment orientation and solids]).

The evaporating solvent can however produce strong turbulences within the wet film – so-called Bénard cells are formed. This applies especially to slow solvent evaporation which has a negative impact on the way of pigment orientation (cloud formation by close ordering, flocculation). In this context, use of components which lose solvent quickly (e.g., cellulose acetate butyrate) and additives which hold pigment particles in dispersion can help. The positive effects of wax dispersions as spacers are among those recorded in the literature [1].

The correlations discussed above also hold in principle for printing applications. However, there is also the effect of the print carrier (e.g. paper), which (more or less) serves as an absorbing medium for the solvent. The smaller the ink penetration of the print carrier, the better the pigment orientation and therefore the metallic effect – the drying time is correspondingly longer.

Flow lines in plastics applications

During injection moulding of metallic pigmented plastics, the pigments orient themselves along the flow lines of the polymer melt. In general this provides optimal orientation. When different levels of orientation meet, so-called flow lines occur in the injection moulded parts (cf. p. 186 [Part IV, Chapter 3: Orientation problems in injection moulding]. Unlike coatings applications, differentiation is not made between leafing and non-leafing effects because of the high application viscosity and low orientation characteristics of the metallic effect pigments.

Optical properties

The metallic effect perceived by the human eye when looking at applications of metallic pigments is a combination of directional reflection of the incident light from the metallic effect pigment surface and light scattering at the edges and unevenness on the metal pigment surface (Figure II-2.7).

Influencing factors

In order to optimise the metallic effect it is necessary to raise the amount of directional reflection from the platelet surface as far as possible and at the same time to minimise scattering at the edges and irregular areas on the metal flake surface.

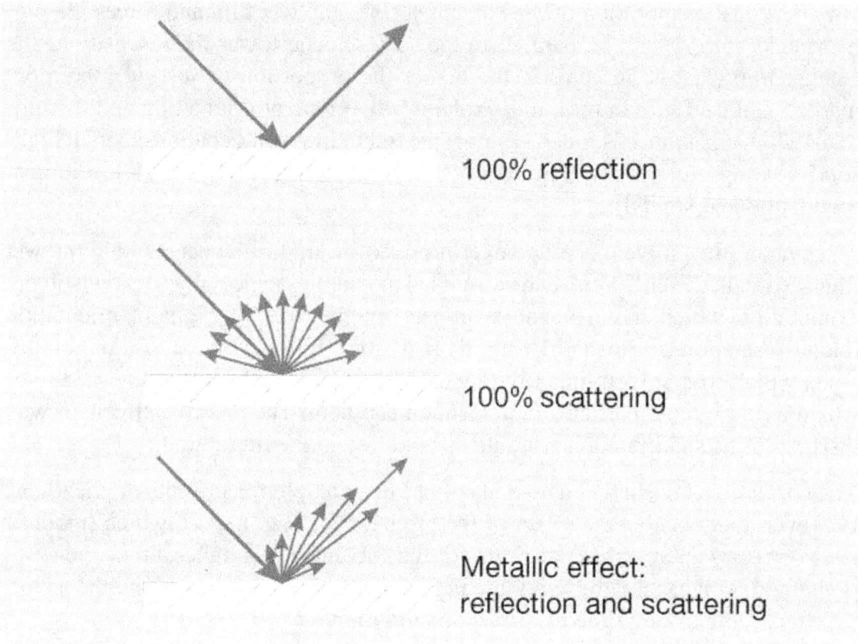

Figure II-2.7: Principle of metallic effect

The optical impression depends mainly on:

- type of metal (aluminium, brass, etc.)
- wetting behaviour of the pigment (leafing or non-leafing)
- surface smoothness of the pigment particles
- particle size and particle size distribution
- form factor (ratio of diameter to thickness) of the flakes
- type of film boundary (air or binder)

Light scattering and random reflection are not only encountered at pigment edges but also at any imperfection on the pigment surface.

Metallic effect pigments are exposed during usual methods of manufacture to strong mechanical forces which often leads to uneven particle thickness and even more to surface defects. Only the expensive PVD process (*Physical Vapour Deposition* (cf. Part III, Chapter. 1)) produces perfect pigment surfaces. The absence of surface defects results in a great amount of directional reflection (cf. p. 53, Figure III-1.4 [Part III, Chapter 1]).

Individual characteristics of the metallic effect on non-applied pigment

It is not easy, without knowledge about the orientation of the particles, to describe and measure the visual metallic effect, as it is a combination of numerous individual effects:

- hue
- whiteness
- brightness, brilliance
- tinting strength or hiding power
- sparkle-effect

Whiteness

Whiteness describes how light a metallic effect pigment appears to an observer and this can be assessed by the L* value of the CIELAB-system (cf. p. 47). The term whiteness characterises the light-dark impression of a metallic effect pigment and must be distinguished from the term metallic brilliance.

Metallic brilliance

"Metallic brilliance" describes the degree of reflection of a pigment and how metallic an effect pigment appears to the human eye. Pigments are brilliant when they reflect a large amount of light and there is little random scattering or absorption causing greyness. In general, metallic effect pigments are the more brilliant the coarser the grade, the more perfect the pigment surfaces and the narrower the particle size distribution. If pigments are very fine, irregularly shaped and have a broad particle size distribution, the colour shifts towards a grey and dull appearance ("dirty grey"). Production of very fine light pigments with high metallic brilliance is a continuous challenge for the manufacturers of effect pigments.

Sparkle-effect

Coarse metallic effect pigments with average particle size greater than 25 μm are referred to as *sparkle or sparkling grades*. The human eye can see and differentiate the reflective tiny mirrors of individual flakes. Such grades are particularly popular in Asia, but also in America and in markets which use American coatings technology: Coarse metallic effect pigments produce attractive effects in the high-solid coatings technology which is typical in these regions. The bright, brilliant glitter appearance compensates for the relatively poor level of orientation of the metal particles in high-solid coatings.

The following parameters of aluminium pigments affect gloss of automotive finishes:

- distinctiveness of image (DOI)
- flop
- colour saturation

1. Distinctiveness of image (DOI)

The DOI value *(distinctiveness of image)* was originally used in the USA for assessment of single-coat paints but is nowadays also used for two-coat paints. The DOI value is determined by measuring the gloss close to the reflection angle. This can be done visually in a glow-box or more objectively by accurate measurement. The values obtained are relative numbers (0 to 100). The higher the DOI value, the clearer the image of high-contrast objects when reflected in the coated surface. The DOI value is affected by the particle size and particle size distribution of the pigment, its compatibility with binders and general flow properties of the clear coat. This term must not be confused with the gloss (see below).

2. Flop

The flop (also known as *Travel or Two-Tone*) is probably the most commonly known optical property of metallic effect pigments and is often simply called the metallic effect. This is understood as the alteration of the brightness as a function of observation angle. This change in brightness can easily be determined visually, observing the test panel on face and after that at oblique angles. In reality, the flop is evaluated by inspection at different angles, in some cases also a change in colour can be observed (colour-flop). The colour-flop can only be found at pearlescent pigments and in a few cases of advanced effect pigment technology.

3. Colour saturation

Colour saturation is the ability of a metallic effect pigment to cover the colour of a coating within polychromatic applications. There is a direct relationship with the opacity (hiding power) of the metallic effect pigment (see below). In general the colour saturation is stronger the finer the metallic effect pigment or the wider the particle size distribution (see below).

Measurement of the metallic effect (flop)

A well-known method for determining the flop of a coating is by measuring the reflected light with a goniophotometer (Figure II-2.8). The coating is illuminated at 45° angle by monochromatic light and the reflected light is measured at several angles between 0° and 90° (cf. p. 46).

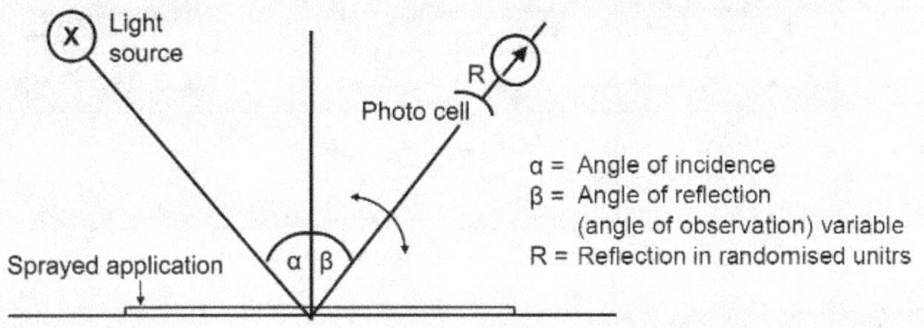

Figure II-2.8: Measurement of metallic effect with goniophotometer

Measurement of the reflected light at various angles shows that there is a clear peak at one particular angle. As the angle of incidence is 45°, the strongest reflection is also at this angle. This highest light reflection corresponds to the brightest visual impression as seen on face; all other values correspond to the darker impressions on flop. Nowadays goniophotometers have been replaced by colour measuring devices which produce more detailed data (multi-angle spectrophotometers). The principals of the angle dependant measurement in goniophotometers differ fundamentally from modern multi-angle spectrophotometers, in which the angle is defined relative to the reflection angle [2]. An angle of 25° on a spectrophotometer means that the light reflection has been determined at 25° anti-clockwise from the reflected beam.

In quality control at least two (or better three) angles should be used to characterise the metallic appearance of a metallic effect pigment. An angle of 25° [= L*(25°)] has become accepted to describe the visual effect on face. Often angles closer to the reflected light ("nearer to the gloss") are used. If for example, measurements are taken at an angle of 15° [= L*(15°)], the accuracy of the measurement will be less as the measurement is "nearer the gloss". Other angles used are 110° [= L*(110°)] and 70°, which measure the darker appearance on flop. Besides an intermediate angle around 45° is common [= L*(45°)].

From three colour measurement values the flop index can be mathematically calculated using a formula recommended by DuPont. This allows pigments to be characterised and compared with each other:

Flop-Index = 2.69 * (a/b)

with a = [L*(15°) – L*(110°)]1.11 and b = [L*(45°)]0.85

Table II-2.4 shows flop data for different aluminium flakes in a defined coating system at equal pigmentation level, applied under identical conditions [3]. The

Product	Fineness	Flop-Index	Flop properties
STAPA® Metallic 801	fine	11.9	low
STAPA® Metallux® 214	coarse	18.7	strong

Table II-2.4: Comparison of flop-indices

flop-values (flop indices) differ because of, for example, different particle orientation and different light scattering due to the flake morphology [4]:

Hiding power or opacity

The term hiding power (or opacity) describes the ability of a pigment to cover a coloured substrate. The coarser the effect pigment, the lower the hiding power. Fine pigments with small average particle size show only a modest, metallic appearance but they have the advantage of an excellent hiding power. In plastic coatings e.g. for mobile phones, audio-visual equipment and toys, where cost limitations restrict the film thickness, the good hiding power of fine metallic effect pigments is a major advantage. The reason for the excellent hiding power of fine metallic effect pigments is the large number of small, tiny platelets – this results in the large specific surface area of these products.

Gloss

In general, increasing fineness of the metallic effect pigment and smaller sieving mesh during the manufacturing process results in increased gloss (the total of all light reflections), i.e. the distinctiveness of image increases. The coarser fraction of the pigment particles is removed from the particle size distribution and cannot disturb the degree of gloss any more. The data in Table II-2.5 clearly display the correlation discussed above between particle size, gloss and metallic effect for various fine pigments for bronzing, gravure- and flexographic-printing, and offset ink products:

Application / Properties	Bronzing	Gravure/flexo ink	Offset
Particle size	35 to 40 µm	10 to 15 µm	3 to 5 µm
Gloss (60° angle)	63%	89%	103%
Metallic effect	excellent	very good	good

Table II-2.5: Optical properties as a function of particle size

Pigments with larger average particle size between 8 µm and 200 µm are used in the plastics field. In coating applications a typical range for particle size between 5 µm and 45 µm is used.

Particle size, particle size distribution, flake morphology

During manufacturing and grinding, the isometric metal granules are converted in the ball mill into extremely thin platelet-shaped metallic effect pigments with a typical particle thickness in the range of 0.1 µm to 0.3 µm. The flakes do not have a single, exactly defined particle size (diameter) but have an approximately Gaussian distribution: a limited number of fine particles, a large number of medium-sized particles and again noticeably less coarse platelets (Figure II-2.9).

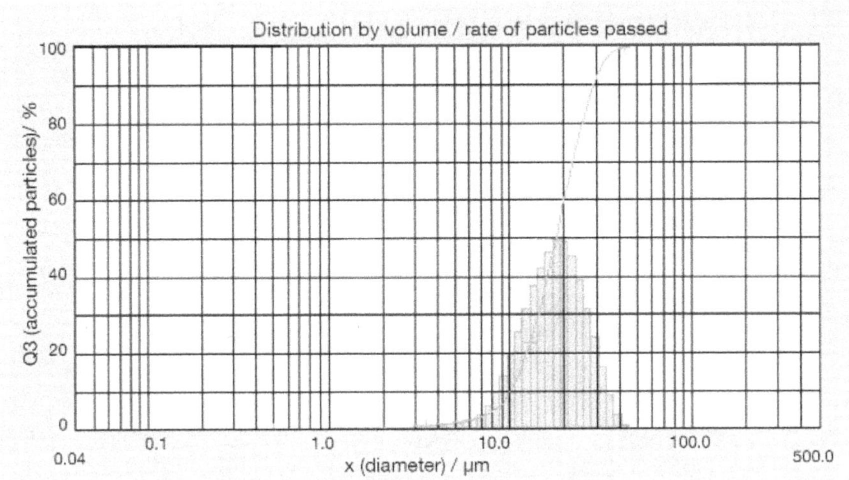

Figure II-2.9: Cilas plot

Optical results

Particle size, particle size distribution and particle shape have a direct influence on the optical appearance of metallic effect pigments. As the optical properties of the coating depend mainly on the reflection of the incident light from the pigment surface and scattering from the pigment edges, the metallic effect depends mainly on the ratio of the surface area to the edges of the pigment. If this ratio is high a bright metallic appearance is observed. The surface/edge ratio is determined by particle size, particle size distribution and particle shape and is affected by the manufacturing conditions.

Average particle size and particle size distribution

Average particle size: d_{50}

The coarser the pigments, that is the larger the average particle size, the higher is the amount of reflecting surface to scattering edges – the metallic effect is bright and brilliant. For the coating system already described and under defined spray conditions the following data apply [3]:

STAPA® Metallic 801 [4]: fine (d_{50} = 14 µm)
L*(15°) = 128.0; L* (45°) = 64.7; L*(110°) = 34.9
Flop-Index: 11.9
Character: grey effect with low flop

STAPA® Metallux 214 [4]: coarse (d_{50} = 37 µm)
L*(15°) = 149.8; L*(45°) = 53.1; L*(110°) = 29.4
Flop-Index: 18.7
Character: bright, metallic brilliance with excellent flop

The scanning electron microscope image of metal flakes (gold bronze pigments with different particle size, Figure II-2.10) illustrates this.

Figure II-2.10: SEM image of a Gold Bronze Pigment

Particle size distribution

Metallic effect pigments obtained by ball milling do not have a discrete particle size. Rather they have an approximately Gaussian distribution (cf. Figure II-2.9). This means that products with the same average particle size but different particle size distribution create a very different optical impression. Pigments with a broad distribution have next to a certain amount of coarse particles also a high proportion of small particles with an extremely high amount of edges and corners for a given size of surface. The ratio surface/edge is low resulting in moderate optical properties: brilliance is low and flop is poor. Metallic effect pigments with a narrow particle size distribution have only a small proportion of fine particles. The surface/edge ratio is high and consequently these products are metallic bright and have good flop. The following values were determined in the coating system already discussed [3] for two types of aluminium flakes with the same average particle size (d_{50} = 17 μm):

STAPA® Metallic 601 [4]: broad particle size distribution
L*(15°) = 134.2; L* (45°) = 62.7; L*(110°) = 34.0
Flop-Index: 13.3
Characteristics: fine-particle, grey effect with low flop

STAPA® Metallux 9157 [4]: narrow particle size distribution
L*(15°) = 141.7; L*(45°) = 59.0; L*(110°) = 31.6
Flop-Index: 15.5
Characteristics: fine-particle, bright, metallic brilliant grade with marked flop

Particle shape

In addition to the particle size and its distribution, the flake form also affects the metallic effect. The shape of the particles varies depending on the manufacturing process. They range from irregular formed platelets known as *Cornflakes* (Figure II-2.11) to almost round particles with minimal scattering proportions which are known as *Silver Dollars* (Figure II-2.12). Silver dollars were developed relatively recently and are manufactured by an elaborate process using special, fine granule fractions.

Figure II-2.11: Cornflake shaped aluminium pigment

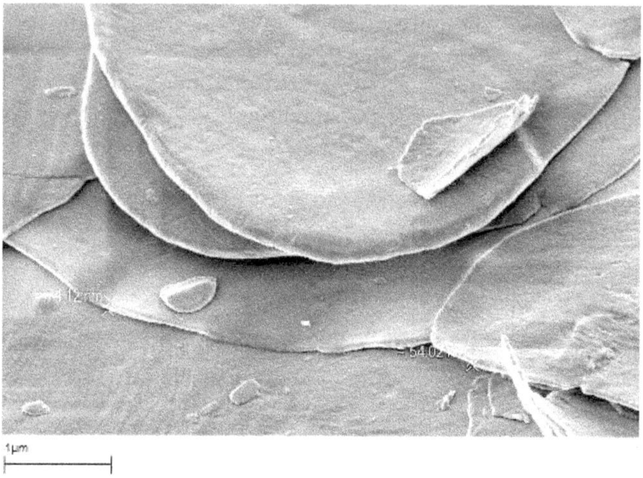

Figure II-2.12: Silver dollar shaped aluminium pigment

Characteristics of silver dollars

In addition to their special flake morphology, silver dollars have a very narrow particle size distribution and this further improves their optical properties. They are the brightest and most brilliant aluminium pigments produced by ball milling currently commercially available. However, their hiding power is only moderately. Application fields range from the automotive sector through high-quality coatings for plastics to high brilliance ink. Typical colour readings for specified conditions of application in the coating already described [3] for pigments with the same d_{50} value are:

STAPA® Metallux 9157 [4]: Cornflake

L*(15°) = 141.7; L*(45°) = 59.0; L*(110°) = 31.6

Flop-Index: 15.5

Characteristics: moderately fine pigment with acceptable optical properties

STAPA® Metallux 2156 [4]: Silver dollar

L*(15°) = 151.1; L*(45°) = 50.8; L*(110°) = 28.7

Flop-Index: 19.8

Characteristics: moderately fine pigment with very good optical properties

As a summary the following can be concluded for ball-milled products: The coarser a pigment and the rounder the shape of the flakes, the higher the amount of directional reflection (and consequently metallic brilliance, brightness and flop)

and the lower the colour saturation in polychromatic mixtures. Bright, brilliant metallic effect pigments with high flop are especially popular in Europe. Ideally they should also simultaneously have high hiding power and high DOI-values. As explained above however, the desired properties (Figure II-2.13) require special manufacturing techniques in some cases.

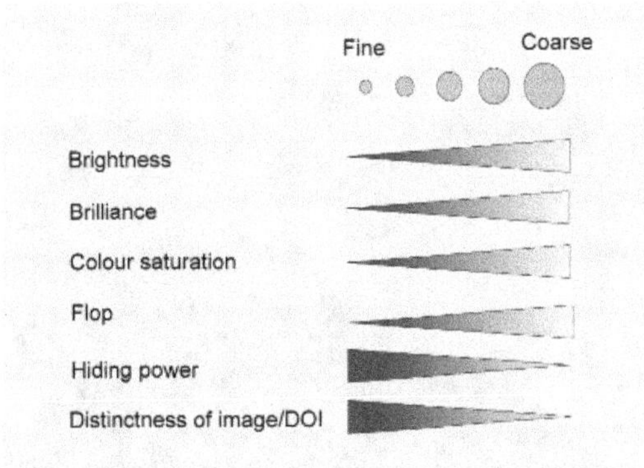

Figure II-2.13: Effects of particle size

Mechanical stability

One major requirement for an optimum in effect is proper dispersion of the metallic effect pigment in the binder matrix or even earlier at the slurry stage, when the pigment is dispersed in a solvent. The effect pigment pastes (delivery form) should be mixed with between one to two times the amounts of solvent and then dispersed slowly by careful stirring. In general, only low shear forces are necessary to convert the effect pigment into the "monodisperse form". Because of their shape and structure, the pigment particles are exceptionally sensitive to high mechanical shear forces. Stirring equipment should therefore be used which does not damage the individual particles (e.g., low-speed butterfly stirrers). The use of roller mills, kneaders, mixers or ball mills commonly used for dispersion of colour pigments damages effect pigments.

Pigment slurry

The preparation of pigment slurries involves the smooth dispersion of pigment powders, pastes or concentrates in suitable solvents. As a second step, these slurries are transferred into coatings or printing binder systems. For the slurry, stirring for about 20 minutes at a maximum peripheral speed of 4 m/s is recommended. The amount of mechanical stress during the dispersion of the pigment slurry can be reduced by allowing the pigment/solvent mixture to soak in advance. This allows the solvent to penetrate into the spaces in-between the individual flakes and subsequently leads to better dispersion. Metallic effect pigments can be partially deformed and damaged by high shear forces (Figure II-2.14). This leads to a significant reduction of the metallic effect.

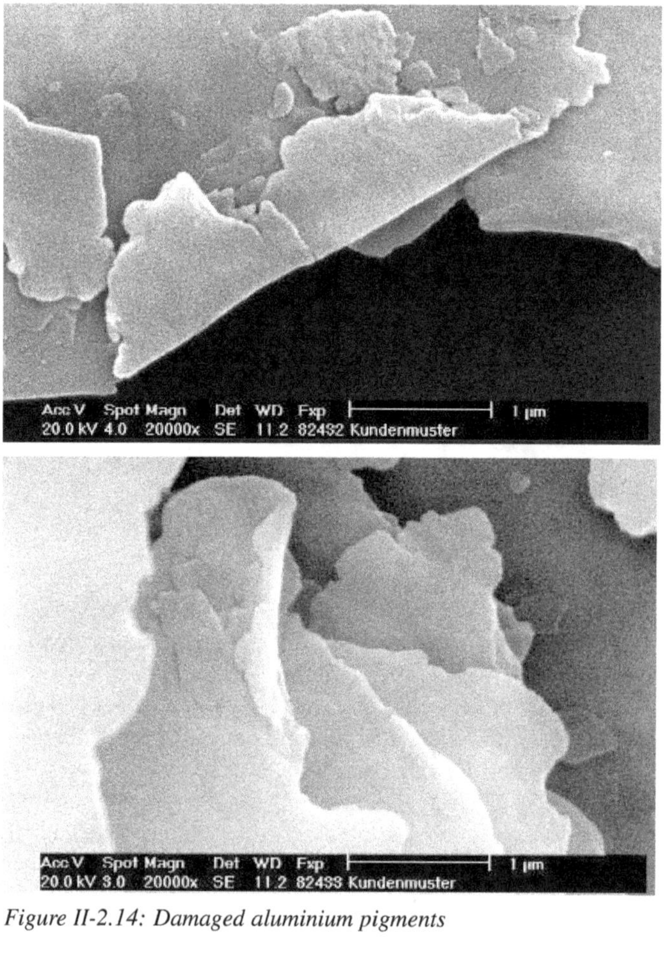

Figure II-2.14: Damaged aluminium pigments

In water based systems, excessive shear force can damage the stabilising protective layer around the aluminium pigment. As a result of this, gassing can be produced; this is dealt with in more detail in section 1.1.2. in Part IV.

Shear-resistant pigment grades

Thicker aluminium flakes, known as *non-degrading flakes* or, *commercially*, as Tufflakes® are significantly more shear stable. A parameter for the classification of these flakes is the *form factor*, which is defined as ratio of the average particle size (obtained by laser measurement of the particle size distribution) to the particle thickness (determined from density and spreading). Typical values are about 100 to 150 for leafing and non-leafing pigments. For non-leafing silver dollars the form factor is between 50 and 70. With aluminium flakes specially manufactured to be shear stable, the value is about 30. In the automotive industry, metallic paints are usually pumped around for long periods of time in circulation lines and internal gear pumps can damage the metallic effect pigments. In addition, high shear forces, which can also cause damage, often occur at pressure regulating valves between the circulation lines and spraying equipment. Mechanical stability is particularly necessary when metallic effect pigments are used in circulation lines with "aggressive" pump systems.

Test methods

Shear stability is usually determined by colour measurements in a given system before and after exposure to shear forces. Comparative values are given in Table II-2.6.

L* (20°)	Stability	Type
< 5	non-degrading, durable	usually Tufflakes®
5 to 10	medium-degrading, semi-durable	usually silver dollars
> 10	degrading, not stable	usually cornflakes

Table II-2.6: Shear stability of different pigment grades

The shear stability is tested with a defined amount of pigmented basecoat exposed to mixing at about 20,000 rpm over a specific time (e.g. 15 minutes) in a water-cooled mixer. The blade of the mixer has very sharp edges similar to those of a kitchen mixer. In addition to this test, the *Turrax-Test* (named after the mixing equipment manufacturer), and many other tests with special equipment have become established, which simulate pigment behaviour in circulation lines of large scale coating lines.

Chemical stability

Resistance

Aluminium is an amphoteric metal which is reactive with both acids and bases. In many coatings the thin oxide layer combined with the fatty acid layer is sufficient to protect the aluminium pigment against aggressive agents. In practice, non-leafing pigments are particularly stable, as they are surrounded by a binder matrix which provides additional protection. Generally speaking, the chemical stability improves with increasing purity of the aluminium pigment. Standard aluminium pigments have a purity greater than 99.7%. High purity, chemically more stable pigments are known as R- (or AR-) grades and have a purity of greater than 99.9% (R stands for *resistant* and AR for *acid resistant*). High purity pigments offer advantages in coatings which contain aggressive components, for example binders with a high acid values. The metal purity has scarcely any effect on the reflection properties of the pigment, so that comparable optical properties are obtained for similar particle size and particle size distribution.

The standard and high-purity grades are both produced by the already described Hall wet-grinding process using aluminium granules of the appropriate purity. If the pigments must have a stability higher than obtainable with high-purity grades, appropriately coated grades must be used (see below).

In spite of the well-known reactivity of metallic effect pigments with their large specific surface area, the pigment industry has successfully introduced surface treatments to manufactured products which are essentially inert while still maintaining their gloss. Modern metallic effect pigments used with an appropriate binder system generally fulfil the highest demands in respect of fastness.

Improved chemical resistance by pigment surface coating

Water, organic solvents, binders with acid groups or solvents or binders having acid functionality (e.g. chlorinated hydrocarbons) can chemically react with metallic effect pigments. With aluminium pigments this shows up as greying or formation of hydrogen gas in the pigmented systems. Aluminium pigments for powder coatings and plastics can be stabilised by SiO_2-coating (cf. Part IV, Chapter 1.1.3 and Chapter 3). Correspondingly stabilised aluminium pigments for water-based systems are obtained by using suitable additives or by coating with inorganic compounds (cf. Part IV, Chapter 1.1.2). Coating with polymers produces chemically resistant grades which are mainly used as electrically insulating products in the electronics industry.

Chemical reactions with gold bronze pigments

Chemical reactions involving copper/zinc pigments are complex. Depending on the actual conditions, either de-zincing can occur or both components of the alloy can be released into solution at the same time. This mainly shows up as a greenish coloration of the binder or the coating. The corresponding heavy metal cations can catalyse the polymerisation or polycondensation of the binder, resulting in thickening or gelling of the system. Similarly, if alkaline components are present in the formulation, under certain circumstances, complex formation can occur with copper and copper/zinc pigments. This type of reaction is generally accelerated through presence of water (or moisture) so that care must be taken not only to use neutral systems with the lowest possible acid value but also to avoid moisture. Mechanical damage to the pigment surface also increases reactivity. From a chemical point of view, the gold bronze pigments undergo oxidation according to the following partial equation producing copper salts (oxide, sulphide, sulphate, etc.) which are coloured and subsequently produce the discoloration already mentioned:

$$2\ Cu/Zn + O_2 + 4\ H^+ \longrightarrow 2\ Cu^{2+}/Zn^{2+} + 2\ H_2O$$

The free copper and zinc ions can now participate further in chemical reactions with other components present in the system or can act as catalysts for other reactions. The reaction of bronze pigments with nitrocellulose has been known for a long time. With addition of heat this reaction liberates nitric oxide and produces thickening of the system. The mechanism involves catalytic complex ion forming of the NC polymer chain by free copper ions.

Gold bronze pigments in water-based systems

From what has been said previously it is necessary that gold bronze pigments for use in water-based systems must also be properly protected respectively coated. One possibility is to stabilise gold bronze pigments against possible reaction with water using hydrophobic additives. This type of post-treated product is available commercially, for example, as a water-based pigment concentrate under the name Rotovario® aqua (manufactured by *Eckart*) [4]. The main fields of application are water-based printing inks and also water-based decorative and industrial coatings.

Overview of gold bronze pigments

Table II-2.7 contains a general overview of gold bronze pigments with important characteristic values and fields of application.

Zinc flakes

Because of its position in the electrochemical period of elements, zinc has a certain potential for reaction in water-based coatings systems and this requires raw materials with low functionality to be used in formulations. Post-treatment with additives or encapsulation of the zinc pigments is not an option as the corrosion protective effect would be lost.

Heat stability

Aluminium pigments melt approximately at 650 °C. As this temperature is not reached even in applications with very high processing temperatures (such as coil coating, powder coatings and plastics) aluminium pigments can be safely used in these classic fields of application. With higher temperature exposures, such as

Type	Average particle size in µm	Upper limit of particle size determined by sieving* [µm]	Applications
Leafing e.g. STANDART® Lack for waterbased coatings e.g. Rotovario aqua, partly Dorolan®	7 to 45	< 45 < 71 < 160	decorative paints, wood coatings, spray cans, DIY
Non-leafing e.g. STANDART® Resist for waterbased coatings e.g. Rotovario aqua, partly Dorolan®	7 to 45	< 45 < 71 < 100	stoving enamels (decorative), 1-coat paints for plastics, wood, etc.

* sieving with various mesh sizes

Table II-2.7: Gold bronze pigments

coatings for exhaust pipes for cars, the binder sinters and the aluminium flakes then melt forming a metallic film.

Gold bronze pigments are more heat sensitive than aluminium pigments. The leafing gold bronze pigments described start altering their colour around 80 °C. This is put to use in fire colouring (cf. p. 18). Gold bronze pigments are unsuitable for plastics, powder coatings or stoving enamels as the processing or curing temperatures greatly exceed 80 °C. However, it has been possible to manufacture coated gold bronze pigments which are more heat resistant. These have already been described as silicate coated STANDART® Resist and Dorolan® gold bronze pigments produced by *Eckart*.

Metallic zinc melts about 420 °C. This temperature limit is above the usual temperature conditions for coatings applications with zinc flakes in air drying systems and in stoving enamels. Problems with heat stability concerning these products are not known.

References

[1] Goldschmidt, Artur et. al., BASF Handbook – Basics of Coating Technology, Hannover 2003, p. 162

[2] Huff, Karl, Visuelle Abmusterung und praktische Farbmessung in der Kunststoffindustrie. Bayer AG 1993, p. 51

[3] Solvent-based, low-solid basecoat with polyester/CAB/melamine (commercial system); 5% pigment loading, 65% pigment paste (full shade, colour silver); spray conditions: SATA-gun, 4 bar spray and control pressure (pneumatic application); nozzle no. 4; 5 passes

[4] Products manufactured by *Eckart*

3 Test methods

Since metallic effect pigments are a pigment class of its own (cf. DIN 55944), they are evaluated in tests that are different to those commonly used for inorganic and organic coloured pigments. Essentially there are two types of tests: those on the pigment itself (i.e. with the powders and pastes) and those on pigmented coating systems.

Standard test methods on powders and pastes

Standard test methods according to DIN 55923/ISO 1247

Current test methods for aluminium pigments (leafing and non-leafing) are compiled in DIN 55923 "Aluminium pigments and aluminium pigment pastes for coating materials – technical specifications". As for safety reasons and for better dispersability, aluminium pigments are frequently supplied as pastes, tests on pastes are of equal importance to those on powders. These tests are also used to a certain degree for gold bronze and zinc flake pigments.

DIN 55923 (corresponds ISO 1247) standardises the following tests:

- water coverage, preferred for leafing pigments
- wet sieving with organic liquids (DIN 53196 and DIN EN 4188/ISO 787)
- amount of organic components soluble in organic solvents
- amount of volatile organic compounds at 105 °C
- amount of metallic impurities
- floating properties (spatula test), preferred for leafing pigments

Determination of specific gravity

Metallic effect pigments require the determination of bulk density in accordance to ISO 3923. Pigment pastes are measured similar to DIN EN ISO 2811-1 in special paste pyknometers.

Special test methods

For unambiguous characterisation of metallic effect pigments and differentiating between products, further tests are needed in addition to those applied as standard routine. The special test methods described below allow the required product quality and consistency to be guaranteed for use in high quality coatings, printing and other applications.

Tests on the pigment

- particle shape
- particle size
- particle size distribution

Testing particle size and particle size distribution is especially important as these determine the metallic effect (see the previous section).

Methods for measurement of particle size

In principle there are a number of different techniques available to determine the particle size and particle size distribution of solids in powder form and in their preparations (e.g. pigment pastes):

- sieve analysis
- sedimentation analysis
- electronic counting (e.g. Coulter Counter)
- light diffraction methods (e.g. laser granulometer)
- image analysis (microscopy)

The significance of the results from most methods is limited by the fact that the shape of metallic effect pigments deviates from spherical.

Sieve analysis

In sieve analysis, the percentage of pigment which remains behind on a sieve of a defined mesh – or respectively passes through – is determined. With platelet shaped particles, it has to be kept in mind, that particles may pass through the rectangular sieve opening diagonally or along their major axis even if the major axis is longer than the sieve cross section. To determine the particle size distribution, sieves with different meshes are used. This is extremely time consuming with the fine pigments used in the graphic arts industry, as the average particle size in this case is around 5 to 10 µm. For this reason sieve analysis is not suitable for routine testing.

Sedimentation analysis

In sedimentation analysis the particle size is determined by the sedimentation rate in a defined liquid. Sedimentation methods are of limited significance for characterising flake shaped pigments as the mathematical equation of sedimentation is only valid for spherical or quasi-spherical particles. Such assumptions lead to incorrect results because of the special sedimentation behaviour of flake shaped particles.

Electronic counting

In this technique, the number of particles of a certain size passing through defined openings of different diameters is determined using electronic equipment. The significance of this technique (e.g. with a Coulter Counter) is also limited by the extreme anisotropy of the particles, so that such methods are also only used in quality control of platelet shaped pigments for special purposes (e.g. determination on the proportion of the very coarse particle fraction).

Laser granulometry

Laser granulometry is currently the most commonly used method for determining the particle size distribution of metallic effect pigments. Although this technique has some restrictions, in principle it fulfils the requirements of speed and reproducibility. It is however not standardised and depends strongly on the configuration of the particular measuring equipment.

Factors affecting measurements with laser granulometry

Sample preparation is particularly important in laser granulometry. Usually the pigments are dispersed in an ultrasonic bath before measurement for a given time. This ensures that the metallic effect pigment is ideally dispersed. Fine particles adhering to larger platelets are removed so that the individual flakes are discrete and can be measured. The length of the pre-treatment and the frequency of the ultrasonic bath should be quoted when stating the results as the energy input affects the degree of dispersion of the pigment. In the extreme case, too long an exposure can mechanically destroy metallic effect pigments and this falsifies the results. The liquid used for dispersion and measurement is also important as different wetting characteristics lead to different degrees of dispersion and hence different results.

Characteristic parameters

The particle size distribution determined by laser granulometry (cf. p. 29 ff) is concisely characterised by three values:

d_{10} as a measure of the fine fraction of the particle size distribution, i.e. 10% of all measured particles are smaller than this value and 90% are larger

d_{50} as a measure of the average particle size distribution, i.e. 50% of all measured particles are smaller than this value and 50% are larger

d_{90} as measure of the coarse fraction of the particle size distribution, i.e. 90% of all measured particles are smaller than this value and 10% are larger

Fundamentals of laser granulometry

Laser granulometry uses diffraction and scattering of light caused by the pigment particles to generate plots of particle size distribution. Coherent laser light falls on a cuvette containing a small amount of properly dispersed metallic effect pigment. Light is both diffracted (Fraunhofer diffraction) and scattered (Mie scattering) in ways which are characteristic of the pigment particle size. The intensity and geometry of the resulting scattering is measured and, using suitable algorithms, a particle size distribution is calculated.

The assumption is made that all particles are ideal spheres. This assumption is not the least bit true for platelet shaped metallic effect pigments and thus the calculated results do not directly correspond to the real distributions. In particular, different assumptions in the mathematical model when evaluating the algorithms mean that the same product will give different results on equipment from different manufacturers and such results can therefore not be compared. However the calculated distribution curves are excellent for monitoring the consistency of a particle size distribution of a particular product and for documenting deviation from the required value.

Microscopic methods of characterisation

Both optical and scanning electron microscopy yield information about particle size, shape and surface nature. An ensemble of minimum several hundred pigment particles is required to analyse representative results for the totality of all pigment particles.

Leafing stability

Leafing is the result of surface incompatibility between the pigment and the coating system in which it is used. The pigment surface is only inadequately wetted which leads to the pigment floating. Certain additives can re-enforce this wetting under certain conditions and disrupt the leafing (cf. p. 19). This means that measurement of leafing stability is important. This measurement reveals how long a leafing pigment will float in a given coating system and when wetting becomes visually apparent. Alternatively the floating characteristics of two leafing pigments can be compared. For measurement, a defined coating or binder system is mixed with the leafing pigment. Immediately a spatula test is made. Then the specimen is stored at room temperature or, more intensively, at 50 °C and further spatula tests are made after one, two, three and seven days. Two possible parameters can be measured: how long the value obtained in the initial spatula test is maintained or, alternatively, how this value decreases over time (Figure II-3.1).

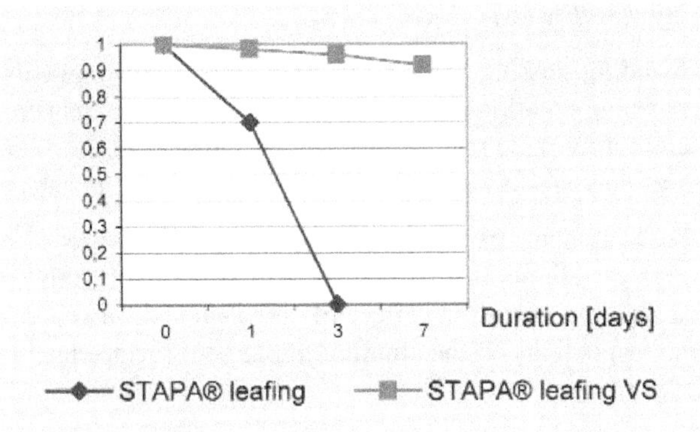

Figure II-3.1: Leafing stability in accordance to DIN 55923 (Spatula test) Vehicle: nitrocellulose system

Stability to shear

For aluminium pigments exposed to high shear forces, as seen in circulation lines in the automotive industry, testing of the mechanical shear stability is important. The relevant test is described in Part II, Chapter 2.

Storage stability

Metallic effect pigments are reactive. On long-term storage they can oxidise or react with moisture, especially in a tropical climate, resulting in alteration of the colour (with gold bronze pigments) or alteration to the particle size distribution. In the latter case, the fine fraction of particles can be lost leading to loss of hiding power and agglomeration of the metallic effect pigment. Storage stability of metallic effect pigments is measured by storage in closed containers at 40 or 50 °C. The stability of the pigments on storage under the given conditions is assessed by wet sieving to DIN 55923 (corresponds ISO 1247), measurement of the particle size distribution by laser granulometry and colour measurement on samples applied to metal test panels or cardboard.

Gassing stability in water-based systems

Aluminium pigments react with water, forming hydrogen gas. It is a must to avoid these reactions in coatings and printing applications and it is therefore important to measure the amount of hydrogen gas produced. The apparatus used for this test is shown in Figure II-3.2. 300 ml of the water-based coating (typically with 5% pigmentation) is stored in a wash flask at 40 °C. Attached to the wash flask is a gas bubble counter with two chambers. The lower chamber contains about 30 ml of water. The volume of gas liberated corresponds to the amount of water displaced from the lower into the upper chamber.

Accelerated gassing stability test

A particularly demanding gassing stability test for aluminium pigments can be carried out at higher pigmentation and in the presence of transparent iron oxide pigments. Transparent iron oxide red and yellow pigments are used in many coatings formulations as they are low cost, lightfast and chemically resistant.

It is known that these pigments can significantly increase the corrosive effect of water-based coatings on aluminium pigments. This is probably related to the large specific area and the desorbing effect of transparent iron oxide (BET > 100 m^2/g). The smallest defects on the aluminium pigment surface lead to strong liberation of hydrogen.

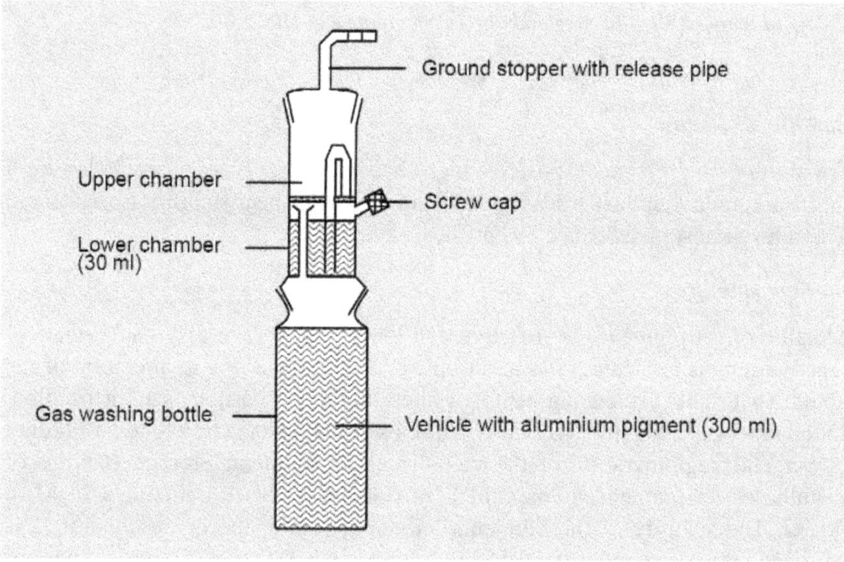

Figure II-3.2: Apparatus for gassing test of aluminium pigments at 40 °C

Different specifications are used for different applications and customised formulations. Acceptable values are between 10 and 15 ml hydrogen in a period ranging from a few days to four weeks after maintaining a test flask in a water bath at 40 °C. A blank, using a wash flask filled with pure water, must always be carried out to correct for variations in the volume of the lower chamber of the gas bubble counter arising from changes in air pressure.

Boiling test

A much quicker method of determining the gassing stability is the boiling test. Because it can be carried out quickly it is suitable for quality control and invol-

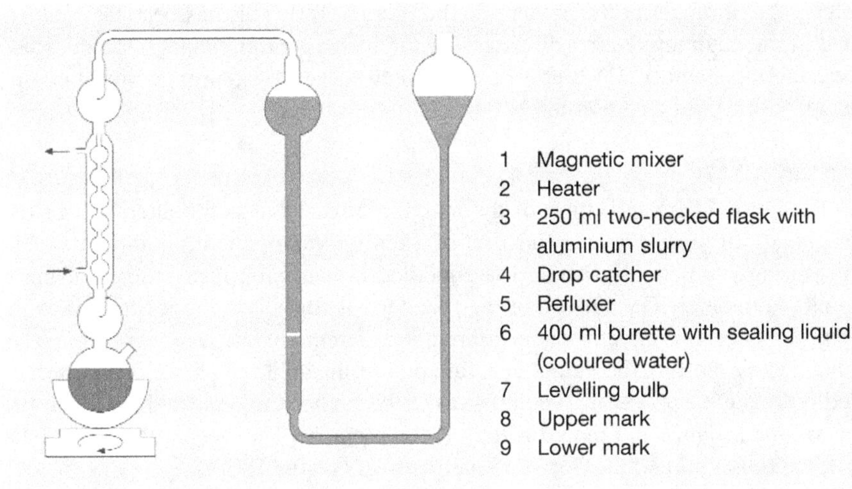

Figure II-3.3: Boiling test

ves determining the time for 1g of aluminium pigment to liberate 400 ml hydrogen from boiling water (Figure II-3.3).

150 ml water is brought to the boil in the round bottomed flask. 1g aluminium pigment (as powder or paste) is dispersed in 10 g of water-miscible solvent and added to the flask. The sealing liquid is adjusted to equal levels in the levelling bulb. Gas formation causes the left sealing liquid level to move. The time (usually in hours) is noted for 400 ml of the sealing fluid to be displaced into the levelling bulb following gassing. It must be noted that, although this test is quick, the reproducibility of the results is not comparable with that of the 40 °C gassing tests.

Tests on pigmented coatings

To test the suitability of a specific pigment for a particular application, further optical examination of the application under practical conditions must be made, such as visual evaluation according DIN 55923 (coresponds ISO 1247). Additionally tests of water condensation to DIN 50017-kk (ISO 6270) and of salt spray to DIN 50021 (ISO 7253) must be carried out.

There are a large number of, sometimes company-specific tests to assess the colour, the fastness and the mechanical stability of a pigmented coating. In these cases the system is always judged as a whole, that is as the composite effect of pigment, binder, solvent, additives, etc. An example of a system test which is acceptable throughout the industry is the Florida-Test. This has become established in the automotive area for assessing the fastness of pigments in outdoor use (weathering stability, UV stability, etc.). In all these cases, a range of test equipment can be used as presented below.

Goniophotometry

Colorimetric characterisation of the metallic effect, that is the alteration of the brightness of a surface as a function of the observation angle, requires variable measuring geometry. An appropriate method is remission measurement using a goniophotometer. The surface of the specimen is illuminated at a constant angle (e.g. 45°) and the reflection is measured at different angles. A reflection curve is obtained by plotting the values as a function of the angle of observation (measured on opaque, sprayed two-coat paints). The higher the typical shoulder in the curve, the brighter and more brilliant is the pigment. The steeper the curve falls off, the more marked the flop or metallic effect (Figure II-3.4).

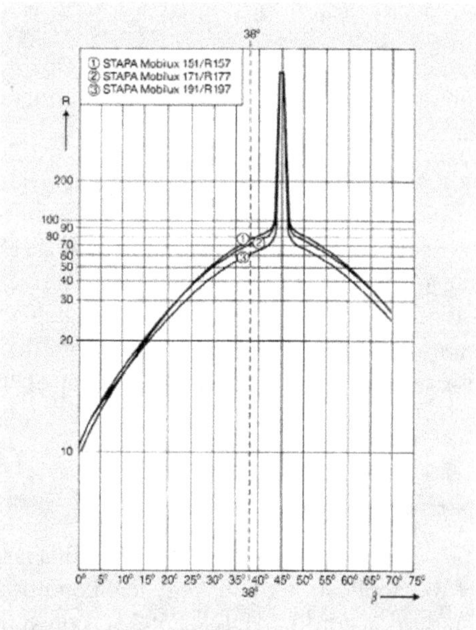

Figure II-3.4: Remission curve (measured with Goniophotometer)

The flop is characterised by the so-called metallic effect (ME-value). It is the ratio of the reflection values at 38° and 0°, with the surface illuminated at 45°:

$$\text{ME-value} = R\,(45°/38°) \times 100 \,/\, R\,(45°/0°)$$

Colorimetry

In recent years, several multi-angle spectrophotometers have become commercially available which allow not only the brightness flop – but also the colour flop – to be measured. Different equipment functions with different geometries but leads to comparable results as measurements are not absolute but are based on a ratio at the difference angle (the angle between the gloss and observation). A schematic of the equipment is illustrated in Part II, Chapter 2.

Fundamentals of colorimetry

Differences in brightness (aluminium pigments) or colour (gold bronze pigments) can be obtained and compared with each other rather precisely using colorimetric values.

As colour differences are visual, subjective impressions and relatively difficult to quantify, there has been a demand to describe them exactly and mathematically. Over a period of time various systems have been developed of which that developed by the "Commission Internationale de l'Eclairage" (CIE) is certainly the most common in the coatings industry.
From experiments with visual colour matching of the three base colours red, green and blue standardised spectral curves have been obtained for normal-sighted observers and these can be mathematically converted to standard tristimulus values. All colours can be displayed in a diagram showing red, green and blue components. This diagram has the "shape of a shoe sole". However, in this model not all colours are displayed in equal distance to each other, which is a great disadvantage. Therefore the CIELab system was introduced, in which brightness (L^*), red-green coloration (a^*) and blue-yellow coloration (b^*) are displayable in equal distance and in the form of a set of three dimensional axes. For non-coloured aluminium pigments only the brightness L^* is meaningful, for the coloured bronze pigments a^* and b^* are also needed.

If metallic effect pigments are to be compared with each other using this method, they must be applied by spraying under identical conditions, i.e. in the same coating formulation (same level of pigmentation) at the same time and under the same conditions of temperature, humidity and air flow. This is important because variations in these parameters can easily cause a change in orientation of the metallic effect pigments and lead to significantly different results. Only careful setting up of identical parameters allows the industry-established batch to batch variation of ΔE^* (25°) approximately two units (sometimes only 1 unit) to be guaranteed.

Brightness measurements by the CIELab system also allow the flop to be estimated for aluminium pigments (cf. p. 26).

Hiding power

In addition to colour measurement, description of the hiding power of an applied coating is also an important optical criterion. This property, which is also known as the *opacity*, indicates the ability to cover a coloured substrate so that the original contrast is cached. The hiding power increases with increasing pigment fineness and therefore with increasing width of the particle size distribution, due to a greater proportion of fine particles.

In the simplest case, the hiding power is determined by comparison of the coating when bar coated or sprayed onto a transparent plastic sheet. Differences in opacity between two specimens can be seen easily if inspected with illumination "from behind" the application.
More exact determination of the hiding power is obtained by applying the pigmented system on pre-printed black/white contrast cards. If the pigment is opaque, the quotient of the colorimetric values over the black and white surfaces is 1.

Gloss measurement on metallic coatings or printing is discussed in Chapter 2 in Part II. As mentioned, in the coatings field, in addition to the usual gloss angle measurements, measurement of the DOI is important.

The variety of application methods for metallic effect pigments requires additional special tests which will be described in the following sections.

Part III Special effect pigments

1 PVD pigments

This chapter presents a special class of aluminium pigments which, despite having been around for about 20 years, are still considered among the "exotic" commercial metallic pigments. This is mainly due to the unique mirror effect which this type of metallic pigment offers thus opening up new vistas in designing surfaces with metallic pigments. The *PVD* process *(Physical-Vapour-Deposition)* by which these pigments are manufactured results in unique characteristics unachievable with conventional pigments, i.e., those which are ground in ball mills (cf. Part II, Chapter 2) except in mixtures.

PVD pigments were first described at the end of the nineteen seventies and at the start of the eighties in two patents [1]. The origin, manufacture, properties and areas of application of these pigments are described in the following sections [2].

Manufacturing process

As mentioned above, the manufacturing process for these pigments is based on PVD technology which is well-known in many other applications. The material (in this case high purity aluminium) is vapourised in a high vacuum and deposited as an extremely thin layer on a substrate.

Manufacture of the pigments described here (s. Figure III-1.1) starts by coating a conventional film (e.g. B. PET, OPP etc.) with a release-coat. This coating process is performed generally by a standardised printing technique such as gravure or flexographic printing. The release-coats are lacquers or binders used in printing industry such as acrylates, cellulose systems, vinyl resins etc. In the next step of the PVD process, the release-coat on the foil is metallised with an extremely thin layer of aluminium. The metallisation process plays a key role in defining the subsequent quality of the pigment since it influences the important pigment properties such as pigment particle thickness and surface structure (see below). In the next step, the foil coated with the release-coat and aluminium is passed through a solvent bath (typical solvents are ketones, esters and alcohols), where the release-coat is detached and the aluminium film is finally released as coarse particles. The resulting dispersion of coarse aluminium fragments, release-coat and solvent is concentrated and washed. The filter cake obtained is dispersed in the desired solvent and broken down using a high speed stirrer into the final particle size which averages approx. 10 to 12 µm. The finished products are available commercially under trade names for example as *METALURE®* [3] in the form of 10% or 20% solvent-based dispersions which are used in solvent-based, water-based and UV curing coatings and printing inks.

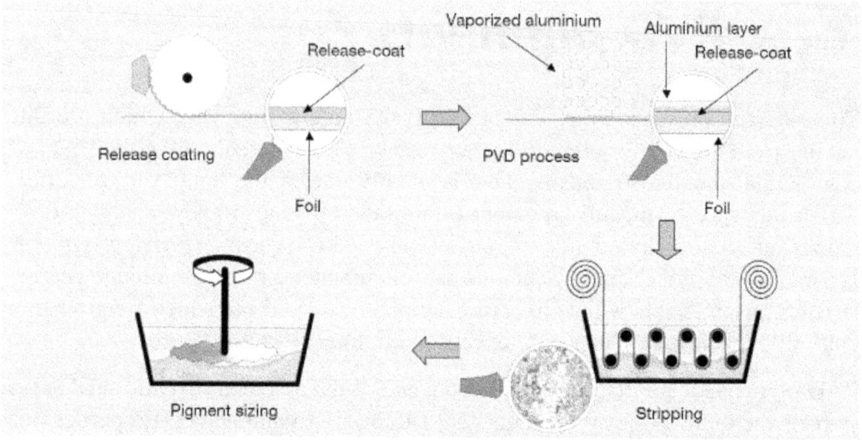

Figure III-1.1: Manufacturing process for PVD pigments

The special characteristics of PVD pigments

Although, as mentioned at the start of this chapter, PVD pigments offer effects which can be achieved with aluminium pigments, no other conventionally produced metallic pigment can reach their optical quality in terms of brilliance and reflecting power. These permit, for example mirror and chromium-like effects to be achieved which can hardly be differentiated from real mirrors or metallised surfaces (e.g. metallised films, plastics and paper). To understand these exceptional properties, a short digression into the principles of the metallic effect and its determining parameters will be made (cf. Part II, Chapter 2).

The metallic effect

The metallic effect perceived by our eyes, when viewing applications of metallic pigments, is a combination of the reflection of incident light on the surface of the pigment and scattering on the edges of the pigment or on surface irregularities (cf. P. 24, Figure II-2.7).

The aim of every manufacturer of metallic pigment must therefore be make the ratio of reflection to scattering as large as possible to ensure maximum brilliance and the most pronounced metallic effect. The discussions on how to maximise this usually consider aspects such as pigment shape (e.g. Cornflake or Silver

Dollar), pigment particle size and particle size distribution (cf. Part II, Chapter 2.). Such considerations can only partially account for the performance of PVD pigments. The parameters mentioned certainly play a decisive role in the high reflecting power of these pigments although closer examination reveals two other factors which are really decisive for the properties described:

- particle thickness
- surface structure of the pigment

Particle thickness

The average thickness of PVD pigments is 30 to 50nm and thus 5 to 10 times less than that of conventional aluminium pigments. Consequently PVD pigments exhibit significantly higher mobility which permits them to orientate themselves within a certain time (e.g. drying time of a solvent-based paint) parallel to the substrate, resulting in maximum directed reflection (the incident light rays are

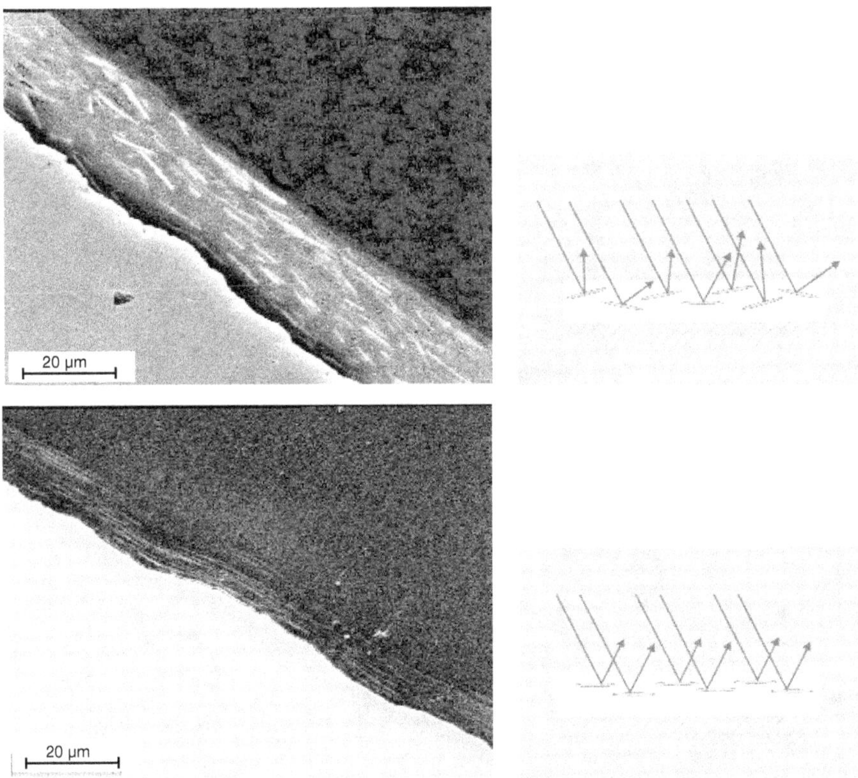

Figure III-1.2: Influence of particle thickness on pigment orientation and reflecting properties

reflected in a single direction). The resulting pigment orientation is thus dependent only on the surface structure of the substrate, the formulation parameters of the PVD-pigmented system (e.g. paint or printing ink) and the application technique and is hardly affected by any lack of pigment mobility.

Pigment orientation

Figure III-1.2 shows the influence of pigment orientation on the reflection of incident light and confirms the difference between PVD pigments and conventional pigments by means of an electron micrograph of a cross section through printed items. The photographs also show the low particle thickness of the PVD pigments which enables each individual particle – comparable to a thin sheet of paper – to fit the substrate. Here again, the important influence of the substrate structure is apparent.

Hiding power

Optimised pigment orientation results not only in the pigment exhibiting a highly directed reflecting power which defines the maximum brilliance (or metallic

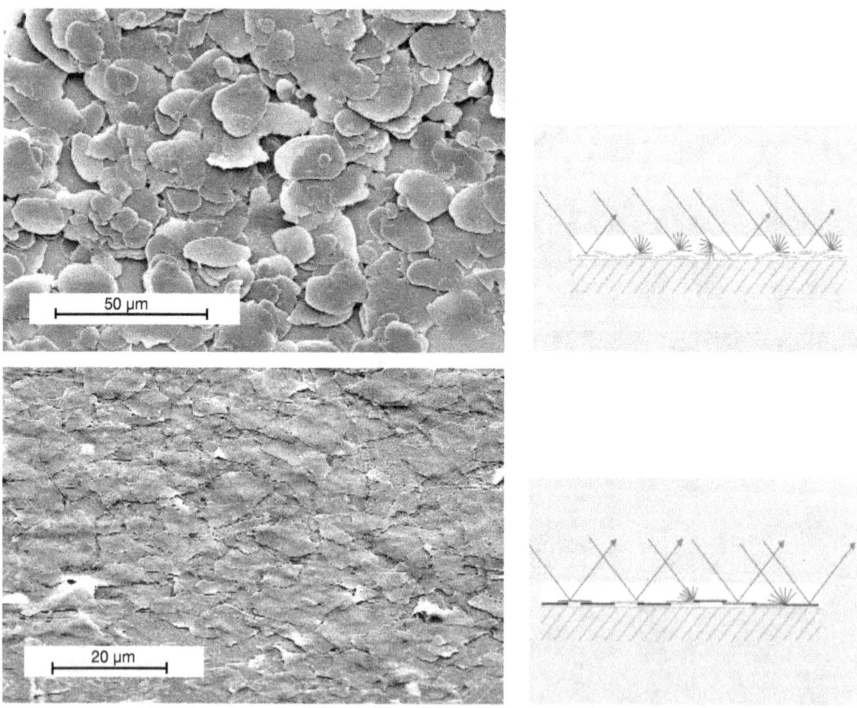

Figure III-1.3: SEM view of printed matter with ball milled aluminium pigments (Silver Dollar) and PVD pigments

effect) and flop effect. Each individual pigment particle contributes to a maximum amount of hiding power since good overlapping of the pigment is assured and thus less substrate is visible than with coatings formulated with conventional pigments. As far as gloss and yield are concerned, PVD-pigmented printing inks and coatings are unsurpassed.

Light scattering

Electron micrographs of applied coatings (Figure III-1.3) show that it is hardly possible to identify individual particles of the PVD-pigments later because of their small thickness whereas, in the coating formulated with conventional pigments, the individual particles are readily apparent. Thus only a small proportion of the light incident on the PVD pigments is scattered and hardly any reduction in reflected light occurs.

Surface structure of PVD pigments

Light scattering and non-directional light reflexion can occur not only at pigment edges but also at defects on the pigment surface. During the manufacturing

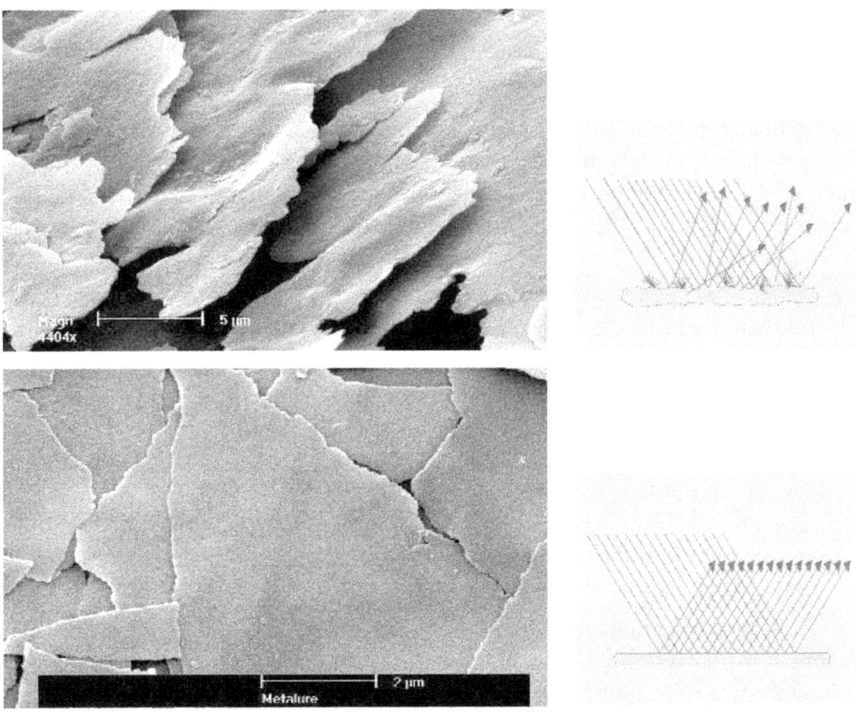

Figure III-1.4: SEM images of surface structure of ball milled aluminium pigments (cornflake) and PVD pigments

process (cf. Part II, Chapter 2), conventional pigments are subjected to external mechanical forces which frequently result in irregular particle thicknesses and particularly in surface defects. Because of the specialist manufacturing process, PVD pigments, in contrast, exhibit an especially smooth, almost defect-free surface and uniform particle thickness. The absence of surface defects results in a high level of directed light (Figure III-1.4).

Non-leafing properties

A final property of PVD pigments should be mentioned, i.e. their outstanding non-leafing characteristics in common coatings systems. This stems from the fact that PVD pigments are manufactured without the usual lubricants (e.g. stearic or oleic acid). This makes them ideal for applications where abrasion resistance, intercoat adhesion and pigmenting power (e.g. generating polychromatic effects) are important.

Areas of application and recommendations

General comments

In the preceding sections of this chapter, the special physical and, above all, geometrical aspects of PVD pigments which result in their remarkable properties were described in detail. To ensure these effects are achieved, factors such as the formulation parameters of PVD pigment based systems (paints, printing inks, etc), application conditions (spraying, brushing, printing) and substrates to which they are applied must be adapted and, if necessary, optimised. To ensure a maximum of directed reflexion, the overall conditions must be selected to ensure the necessary mobility of the pigment and parallel orientation to the substrate.

Formulation

The following aspects should be observed when formulating with PVD pigments:

- the viscosity of the coating system should be as low as possible
- the binder content of the formulation should be as low as possible or the pigment/binder ratio should be as high as possible
- the pigment content should not be too high as this limits the movement of pigment particles
- the applied coating film should be as thin as possible
- the substrate should be smooth and even (e.g. polished metals, high quality films, coated paper, primed plastics)

Unfavourable conditions may be encountered with, for example, high viscosity formulations or systems with a high binder content (e.g. conventional offset or screen printing inks, high solid paint formulations, or with applications involving thick films or on uneven or strongly absorbent substrates (e.g. printing on absorbent paper).

Applications in the graphic arts industry

Although PVD pigments were originally introduced in the coatings industry, their commercial growth in recent years has been mainly in the printing ink industry. In this sector, PVD pigments are used as a cost-effective alternative to metallised printed articles or foil embossing. Suitable adjustment of the printing ink formulation and choice of materials as print substrates enables high brilliancy effects to be achieved and hardly any detectable differences to the alternatives mentioned (cf. Figure III-1.5). PVD pigment-based printing inks are thus widely used in label and packaging industries, especially where a certain part of the printed material should have a high brilliancy metallic finish. PVD pigment-based printing inks can, in such cases, be applied selectively to defined areas of the printed image, whereas in the case of metallised print material, the non-metallised areas of the image must be overprinted with coloured ink or the metallisation removed in a costly process (demetallisation). Because of the unique effects and extremely high costs, PVD pigments seldom compete with conventional metallic pigments in this sector. Further information on the use of PVD pigments in printing inks will be found in Chapter 2, Part IV.

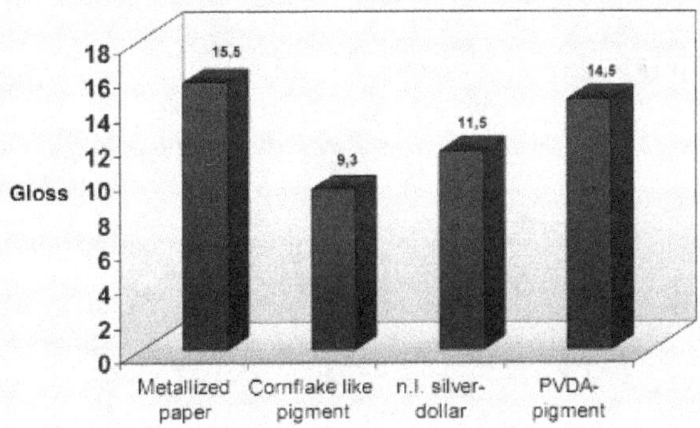

Figure III-1.5: Metallic gloss of print based on PVD pigments compared to that of metallised paper and ball milled aluminium pigments

Applications in the coatings sector

The exceptional brilliance and hiding power and pronounced flop effect of this class of pigments have led to increased interest from the industrial paint and automotive finish sectors in recent years. PVD pigments are now firmly established in coatings industry and are used in numerous applications, in some cases replacing alternative coatings (e.g. metallisation, chromeplating) or conventional pigments. Examples include:

- car interiors
- bicycle finishes
- consumer electronics (housings for audio and video equipment)
- mobile phones
- mirror coating of plastic parts, e.g. for sanitary ware
- to a limited extent painting of automotive wheel rims

Figure III-1.6 shows gloss measurements on finishes based on PVD pigments and conventional metallic pigments.

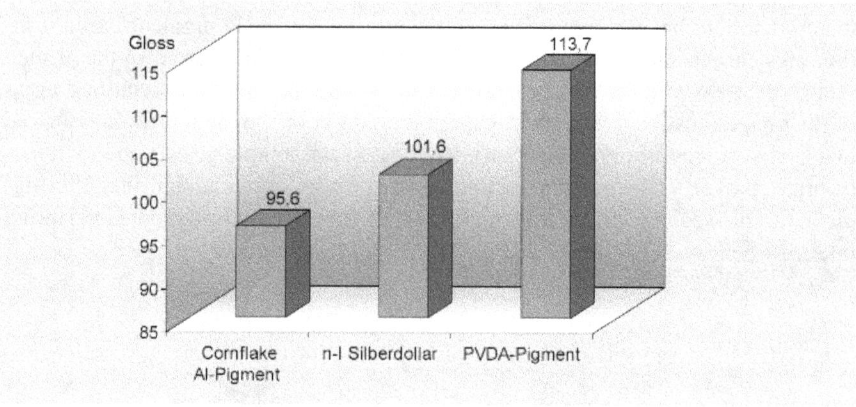

Figure III-1.6: Gloss measurements on finishes with PVD and ball milled pigments

Further information on the use of PVD pigments in the coatings industry will be found in Chapter 1.1.1, Part IV.

References

[1] US Patent 3,949,139 und US Patent 4,321,087

[2] Seubert, J., Fetz, A.: Surface Coatings International Part A. 2001/06,
Seubert, J., Fetz, A.: High Performance Pigments Conference. Miami 1999,
Seubert, J., Fetz, A.: PVD aluminium pigments: Superior brilliance for coatings & graphic arts. *Paint & Coatings Industry*, June 2000

[3] METALURE® is a registered trademark of *Avery Dennison Corp.* Distributed exclusively world-wide by *Eckart GmbH & Co. KG*

Guiding Formulation
2K-Chromeffect with Metalure Rec.-No. 54-09015 ECKART

Component I

1.	Metalure L 55700	14.80	www.eckart.net
2.	CAB 381-2, 20% in Buac 98/100	12.70	www.eastman.com
3.	Lipotin 100	0.07	www.lucas-meyer.com
4.	Uracron CY 474 E, 70% in butyl acetate	1.50	www.dsm.com
5.	Haftharz LTW	0.07	www.degussa.com
stir for 20 min. at slow agitation			
6.	Butyl acetate 98/100	23.50	
7.	Baysilone MA, 0.4% in xylene	1.30	www.borchers.de
8.	Zonyl FSG, 1% in Buac 98/100	2.60	www.dupont.com
9.	Methoxy propyl acetate	33.46	
		90.00	

add pos. 6 to pos. 9 while stirring into a mixture of pos. 1 to 5

Component II

	Desmodur N 75	1.43	www.bayer.com
	Butyl acetate	8.57	
		100.00	

mixing ratio:	90 : 10
spraying viscosity:	12 sec. in DIN 4 cup
diluent:	Methoxy propyl acetate
drying conditions:	air drying or 1 h at 70 °C
pot life:	5 h

Guiding formulation **Chrome-Effect-Paint for metal, water based** **with Hydroshine WS 1001** Rec. No. 56-09002		ECKART
Setalux 6801 AQ-24	28.45	www.resins.akzonobel.com
Water demin. – premix –	7.60	
Butyl glycol	1.85	
Water demin.	9.30	
DMEA (5% in H_2O demin.) – premix –	0.85	
Hydroshine WS 1001	12.45	www.eckart.net
Butyl glycol	5.00	
Setal 6306 SS-60	4.70	www.resins.akzonobel.com
Maprenal MF 900 w, 95% – stir 5 min. –	1.10	www.ucb-group.com
DMEA (5% in H_2O demin.)	3.00	
Water demin.	25.70	
	100.00	

Processing:
First add component b to a, then add c. After that, you add comp. d

Spraying viscosity:	15 sec. in DIN 4 cup
Diluent:	water demin.
Stoving conditions:	30 min at 160 °C

2 Coloured aluminium pigments

Coloured metallic effect pigments are manufactured most simply by using coloured metals. The most common are copper and brass pigments, the colour spectrum of which can be extended by suitable oxidation processes (fire colouring) (cf. p. 18).

Coloured aluminium pigments and interference pigments based on aluminium substrates have been commercially available and growing in market share for over 15 years. They enable novel optical effects to be achieved which considerably exceed the colour space of classic metallics. Aluminium is particularly suitable as a substrate because of its high reflectivity for visible light. Figure III-2.1 gives an overview of the various types of coloured aluminium pigments. The colour effects are originating either by absorption in the pigment itself or by interference effects. Aluminium always serves as a reflector.

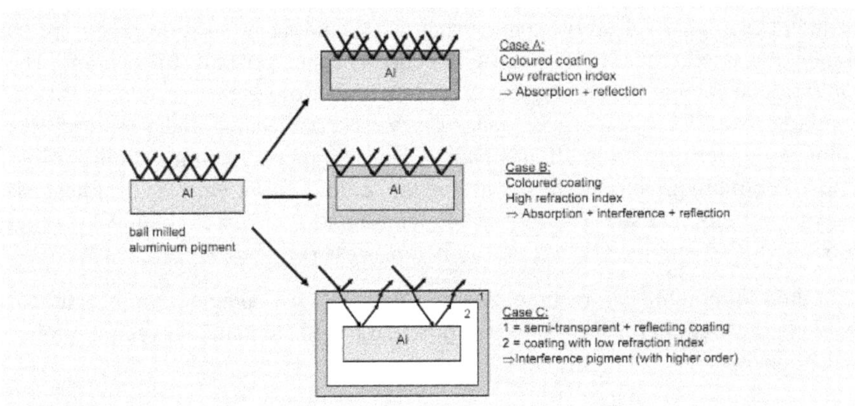

Figure III-2.1 Overview of different grades of coloured aluminium pigments

Case A: low refractive index

In the simplest case (A) the aluminium pigments are coated with a coloured, transparent or semi-transparent layer with a relatively low refractive index. The colour effect is based mainly on absorption. This chapter deals primarily with this type of effect pigment.

Case B: high refractive index

A further possibility (B) consists of coating the aluminium substrate with a coloured, transparent or semi-transparent layer with a high refractive index n (n should be larger than 2). The colour effect stems in this case from a combination of reflection, absorption and interference. Typical examples are iron oxide-coated aluminium pigments in BASF's Paliocrom® range of brilliant, intensely coloured gold and orange shades [1 to 4]. The optical properties of these effect pigments are defined by a combination of the metallic reflection of the base material, the absorption of the iron oxide layer (body colour) and interference effects. The iron oxide layer is applied to the aluminium pigment by a Chemical-Vapor-Deposition Process (CVD) in a fluidised bed reactor.

Case C: interference colours

In the third case (C) the interference colours are generated by a pigment coated with five layers. The aluminium substrate is first coated with a thick colourless layer with a low refractive index (n smaller than 1.8) and subsequently with a highly refractive semi-transparent reflective layer. The low refractive index layer usually consists of MgF_2 [5 to 9] or SiO2 [10 to 13]. The reflective layer consists of metal or highly refractive oxides such as haematite iron oxide. They are so thinly deposited that they are partially reflective and partially transparent. The interference colours of such pigments are of a higher order. Such products are e.g. available from inter alia BASF under the Variocrom® name and from the Flex company under the Chromafflair® name. These pigments exhibit extreme colour flops. The pigment surface layers are manufactured either entirely by Physical-Vapor-Deposition (PVD) [5 to 9] or by a combination of a wet chemical coating process (SiO2) and the CVD fluidised bed process (e.g. Fe_2O_3) [10 to 13].

Coloured aluminium pigments of the first type (A) are obtained almost exclusively via the wet-chemistry route. It is possible to differentiate between the following concepts:

- application of a homogeneous colouring layer
- fixing of coloured pigments in a transparent matrix on the surface

Concept 1: oxidised aluminium pigments

In the case of concept 1, oxidised aluminium pigments under the Aloxal® name (Eckart) have become established commercially. Here the coloured layers are formed by controlled wet oxidation of the suspended aluminium flakes with water [14 to 16]. An aluminium oxide-/hydroxide layer is formed which gives rise to a "champagne coloured" appearance to the aluminium flakes (cf. Figure III-2.2). The colour effect is a combination of absorption and interference phenomena.

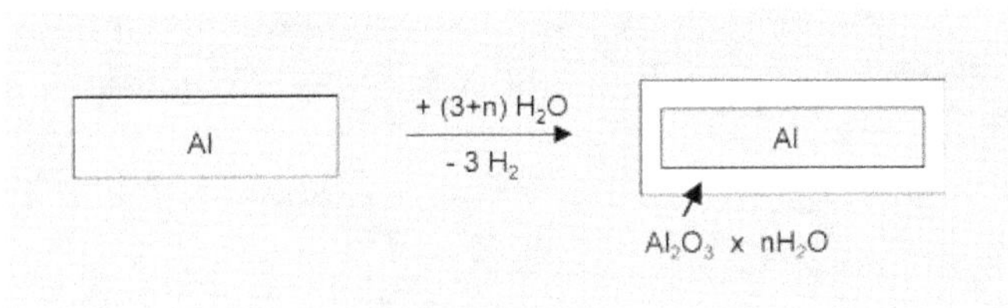

Figure III.-2.2: Cross section of oxidized aluminium pigment

Oxide coating

Besides generating the desired colour, oxide coating has other advantages: It acts as corrosion protection (chemical and gassing stability) and provides increased mechanical strength to the pigment. These pigments represent a counterpart to anodised aluminium surfaces since the oxide layer is formed electrochemically by anodic oxidation. However, since no defined external potential can be applied to dispersed pigment particles by a potentiostat, chemical oxidation is used:

$$Al_{disp.} + (3+n)\,H_2O \longrightarrow Al/(Al_2O_3 \times nH_2O) + 3\,H_2$$

This reaction does not result in the deposition of a coating to aluminium particles in the conventional sense. Rather, a compact crystalline oxide layer grows on the aluminium pigment which is more compact than the more voluminous elemental aluminium oxide. The thickness of the oxide layer is between 70 and 100 nm.

Manufacturing process

The manufacturing process is shown schematically in Figure III-2.3. Aluminium pigments are suspended in a water-miscible solvent in a reaction vessel. After adding the required amounts of the oxidising agent, water, the mixture is heated to the reaction temperature. After a certain induction time, the pigment oxidises with vigorous liberation of hydrogen. Above a certain oxide layer thickness, the typical champagne colour appearance of this pigment is visible. Subsequently the entire contents of the reactor are cooled to room temperature and, after screening out coarse particles outside the correct size range, the pigment is concentrated in a filter press. The filter cake is dried under vacuum and subsequently processed with addition of solvent into the finished paste.

Textured surfaces

Figures III-2.4 and III-2.5 show Scanning–Electron-Micrographs (SEM) and Transmission-Electron-Micrographs (TEM) of oxidised aluminium flakes. In Figure III-2.4B a textured surface of Aloxal® is clearly visible at magnifications

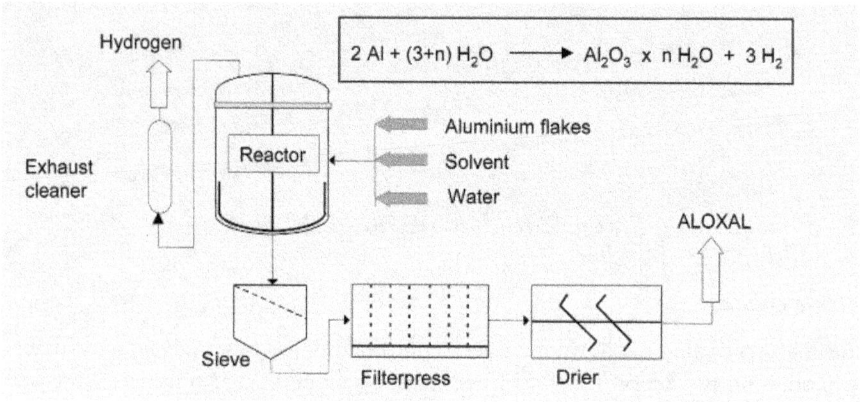

Figure III-2.3: Production process of Aloxal®

greater than 25,000. At 48,000 magnification, (Figure III-2.4C) formation of crystalline structures with dimensions of approx. 50 nm can be seen.

An apparently porous oxide layer approximately 80nm thick is apparent in Figure III-2.5. Despite the partially open structure of the aluminium oxide/hydroxide layer, the pigment is resistant to gassing in water-based paint. During oxidation of the aluminium flakes, a rather dense oxide layer is formed underneath the porous oxide layer. Similar behaviour is known in the case of polished aluminium surfaces which are oxidised in a humid atmosphere at elevated temperatures. Here too, there is an impervious amorphous aluminium oxide layer beneath the

Figure III-2.4 (A to D) SEM Images of Aloxal®

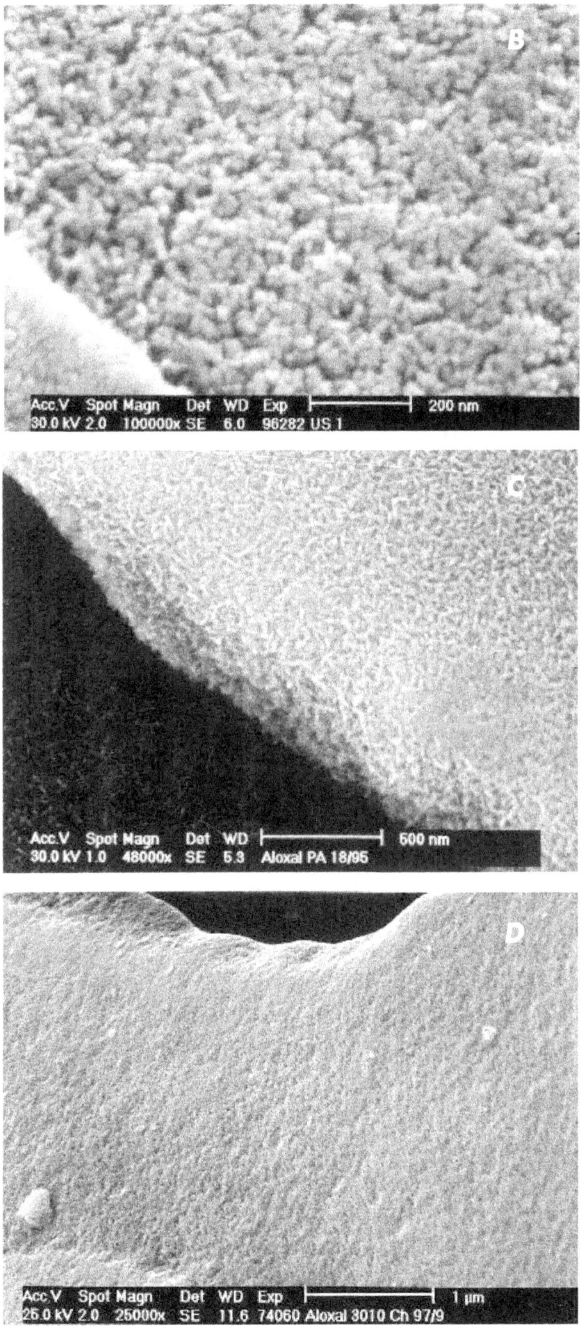
B
Acc.V Spot Magn Det WD Exp 200 nm
30.0 kV 2.0 100000x SE 6.0 96282 US 1
C
Acc.V Spot Magn Det WD 500 nm
30.0 kV 1.0 48000x SE 5.3 Aloxal PA 18/95
D
Acc.V Spot Magn Det WD Exp 1 µm
25.0 kV 2.0 25000x SE 11.6 74060 Aloxal 3010 Ch 97/9

crystalline layer which protects against corrosion [17]. In the case of oxidised pigments, the increase in specific surface area (BET) is due to micro structures which are formed by the liberation of hydrogen gas during the oxidation of the aluminium pigment surface.

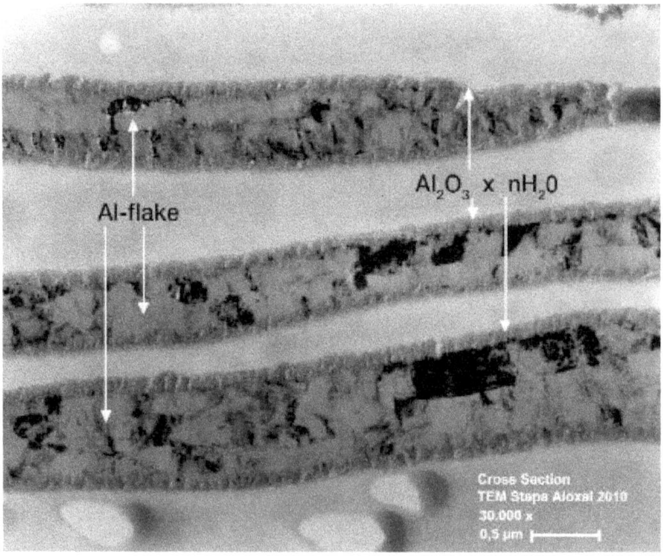

Figure III-2.5: TEM image of cross section of oxidized aluminium flake

Colour definition

Oxidised aluminium flakes appear yellowish close to the gloss angle and increasingly colourless at flat angles of observation. When combined with coloured pigments, a metallic, yet relatively "soft" appearance is obtained which cannot be reproduced with conventional aluminium pigments. It can be deduced from Figure III-2.6 which shows the a*b* coordinates of a blue shade (steel blue) with Aloxal® and a formulation with conventional aluminium flakes, that a wider colour space is covered with the oxidised aluminium pigment.

Property profile

The aluminium oxide/hydroxide layer of oxidised pigments results in good gassing stability. Only in very critical water-based paint systems, additional post-stabilisation with additive treatment of the pigment surface is necessary (cf. p. 86). The solid oxide layer also stabilises the ductile aluminium pigment mechanically against the effect of strong shear forces thus achieving outstanding stability in circulation lines. The weather fastness, solvent resistance, etc. of oxidised alumi-

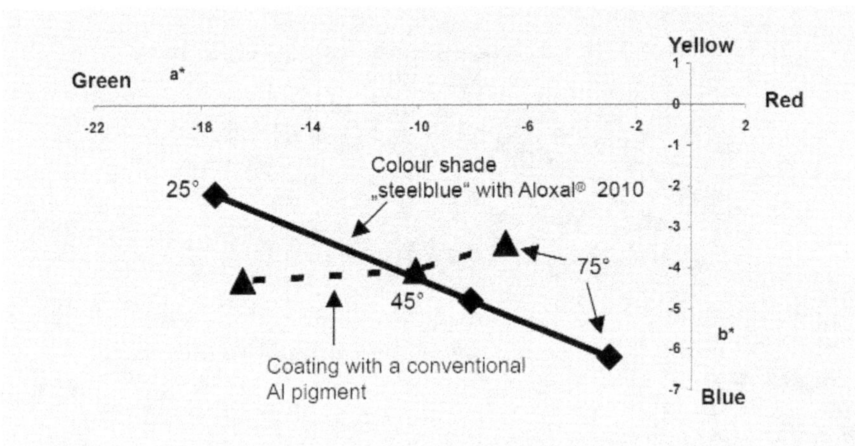

Figure III-2.6: Colouristics of oxidized aluminium pigments
A polychromatic mixture based on Aloxal, compared to a formulation using conventional aluminium pigment. Data from multi-angle spectrophotometer with fixed angle of illumination at 45°

nium pigments are comparable to those of conventional aluminium flakes. This property profile together with the optical properties makes such pigments ideal for automotive applications.

Concept 2: coloured aluminium pigments

In concept 2 colour pigments (CP) are anchored on the surface of the aluminium flakes by embedding in a coating matrix (Figure III-2.7). A typical product is Alucolor® (Eckart). In contrast to concept 1, this method permits all colours to be manufactured. Acrylate [18 to 22] and SiO_2-layers [23 to 26] are reported to have been used as transparent matrices. The silicate coating is formed from tetraethyl silicate using a sol-gel process:

$$Al_{disp.} + FP_{disp.} + Si(OEt)_4 + 2\,H_2O \longrightarrow Al/(SiO_2/FP) + 4\,EtOH$$

The thickness of the SiO_2/colour pigment coating is approx. 200 nm. Weather resistant pigments with a maximum of transparency are generally chosen as colour pigments.

The surface of the coated aluminium flakes can be modified using suitable additives for optimisation of technical pigment properties such as resistance to condensation, pigment orientation in the paint (consequently the optical appearance) and rheology [27 to 28].

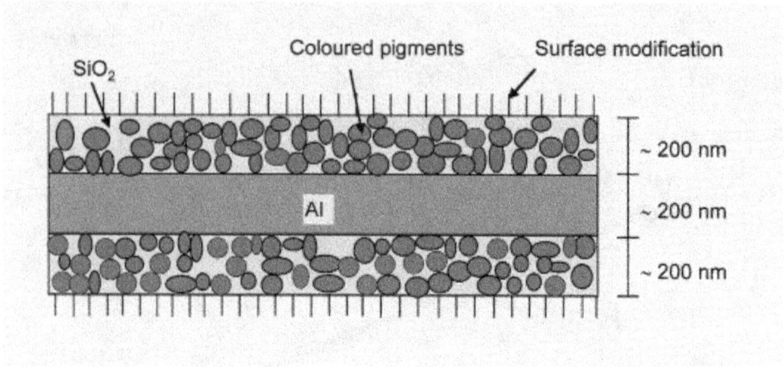

Figure III-2.7: Structure of coloured aluminium pigments (Alucolor®)

Manufacturing process

The manufacture of colour coated aluminium flakes requires coloured pigments which should be present as far as possible as primary particles. To obtain high colour intensity, all the agglomerates in the coloured pigment powder must be completely dispersed [29, 30]. The first step in production process therefore consists of grinding the pigment and stabilising it with additives. The second step involves the wet coating of aluminium flakes by chemical means, e.g. with silicon dioxide, and the encapsulation of the colour pigment. In the presence of aluminium flakes, these sol-gel processes build up a thick layer around the aluminium flakes in which the colour pigments are embedded. The subsequent production steps are analogous to those in the conventional manufacture of aluminium pigments.

Characteristics of coating and surface

The SEM images in Figure III-2.8 show uniform coating of the aluminium flakes with colour pigments. A clear roughness is apparent which comes from the embedded pigment particles in this coating.

The rough surface structures of colour-coated aluminium pigments can now be compared to those of the uncoated starting material by determination of the specific surface (BET method). Result: The specific surface of the coated pigments increases by a factor of about 3.

Figure III-2.9 shows TEM images of a microtome section of Alucolor® at two different magnifications. There is a clearly visible contrast in Figure III-2.9A between the aluminium core and the colour pigment/SiO_2 layer. The coating is uniform and about 200 nm thick. In the boundary areas of the coating (Figure III-2.9B), darker particles with a size between 30 to 50 nm can be spotted, surrounded with lighter areas. These sizes correspond to the primary particles of

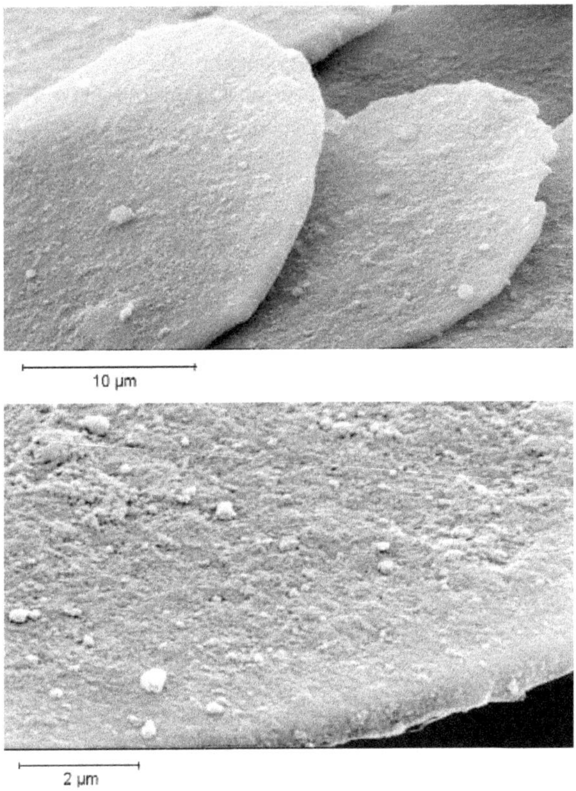

Figure III-2.8: SEM images of Alucolor, with SiO_2 colour pigment

the colour pigment together with their SiO_2 coating. This coating can be considered as nano crystalline, at least partially.

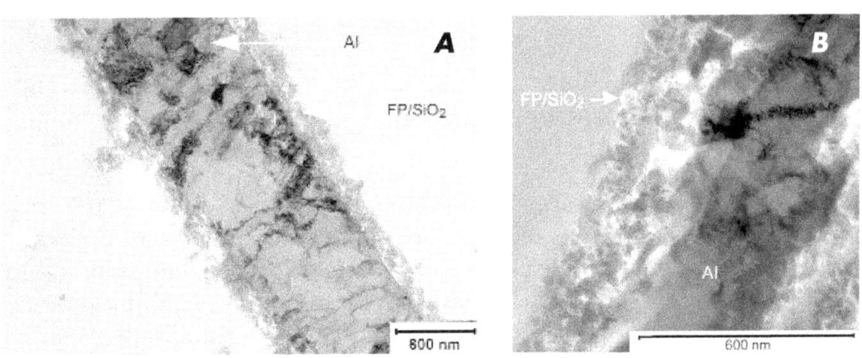

Figure III-2.9 (A to B): TEM images of cross cut of Alucolor®

Pigment content and properties of the colour pigments

The concentration of the colour pigments within the SiO_2 layer is in a range which can hardly be achieved in a paint formulation, even at very high pigment content. The PVC (Pigment-Volume-Concentration) of the colour pigment within the SiO_2 layer ranges from 30% to almost 60%!

The optical properties of the embedded colour pigments are comparable to the typical absorption characteristics of pure colour pigments as transmission spectrometry on microtome sections shows. Despite the high pigment concentration within the SiO_2 layer no significant change in optical properties occurs. These pigments also possess very good gassing stability because of their SiO_2 coating. The colour pigments are so firmly bonded via the SiO_2 matrix to the flake surface that a good stability in circulation lines is achieved (Waring-Blendor-Test).

Colour effect

The main colouristic characteristics of coloured aluminium flakes are extremely intensive chromaticity (Chroma) and high colour purity with slightly reduced metallic brilliance. Table III-2.1 shows clearly the difference in colorimetric values of a coloured pigment and a conventional pigment. Figure III-2.10 illustrates this comparison visually.

Measuring angle	STAPA® Alucolor® PM3503 (2.6% Alucolor, low solid)	STAPA® Metallux® PM2154 (4.0% paste, low solid)
L* 25°	59.4	74.6
L* 45°	30.1	35.8
L* 75°	16.5	19.1
C* 25°	61.7	45.4
C* 45°	38.7	28.7
C* 75°	32.9	23.9
H* 25°	226.9	229.0
H* 45°	251.9	252.2
H* 75°	274.3	271.0

Table III-2.1: Comparison of CIELAB-colour values

Use in plastics

The difference between coloured and conventional aluminium flakes is particularly obvious with coarse grades: as soon as coloured aluminium pigments are seen by the eye as individual particles (Sparkle-Effect, cf. p. 25), they can be clearly differentiated from conventional metallics. This is particularly pronounced in plastics applications where amongst others also coloured glitter is used

Figure III-2.10: Comparison of a ball milled pigment (STAPA® Metallux® 2154) with a coloured pigment (STAPA® Alucolor® PM 3503)

(mean particle size d50 ca. 150 µm) (cf. p. 181). The use of high performance organic pigments makes the coloured aluminium flakes resistant to temperatures of more than 300 °C. This is particularly advantageous for applications in engineering plastics.

Use in powder coatings

In powder coatings, coloured aluminium pigments generally exhibit good electrostatic chargeability. The colour pigment/SiO_2 layer acts as a good dielectric. Up to now, however, mixtures of aluminium flakes with colorants have hardly been used in the compounding of powder coatings. Complete non-leafing, as in liquid paints, is normally not achievable in powder coatings. Therefore in case of conventional aluminium flakes combined with a coloured powder base, the coloured base is partially covered by the aluminium flakes on the surface with their quasi-leafing characteristics. The result is significantly weakening the chroma and resulting in a finish with a cloudy "spotty" appearance, where coloured and silver areas alternate. If however, coloured aluminium flakes are used, an intensive chroma is obtained with a homogeneous metallic appearance and pigment distribution [2].

Conclusion

Aloxal® oxidised aluminium flake is an innovative new aluminium pigment generation. Because of its unique colour, it opens up new interesting possibilities in styling of automotive paints and other effect finishes. These products are economical in use and entail little risk as raw material for coatings applications because they behave similar to conventional aluminium pigments. Similar advantages can also be assigned to Alucolor® coloured aluminium pigments, which permit for the first time the manufacture of powder coatings with excellent depth of colour and opacity.

References

[1] *Ostertag, W.; Mronga, N.; Hauser, W.*: Eisenoxidbeschichtete Aluminiumpigmente.Farbe & Lack, December 1987, p. 973 – 976
[2] BASF, EP 033 457
[3] BASF, EP 338 428
[4] *Mronga, N; Radtke, V.; Baumann, B.*: Paliocrom lustre pigments – their production, properties, and application. 3rd Nürnberg Congress 1995, Paper 18
[5] Flex Products Inc., US 5,059,245
[6] Flex Products Inc., US 5,135,812
[7] Flex Products Inc., US 5,569,535
[8] Flex Products Inc., US 5,571,624
[9] Flex Products Inc., US 5,635,792
[10] BASF, EP 0 571 836
[11] BASF, EP 0 668 329
[12] BASF, EP 0 708 154
[13] *Schmid, R.; Mronga, N.; Radtke, V.; Seeger, O.*: Optisch variable Glanzpigmente. Farbe & Lack, May 1998, p. 44 – 48
[14] Eckart, DE 195 20 312, 1995
[15] *Greiwe, K.*: Coloured aluminium pigments. In: 4th Nürnberg Congress 1997, Paper 6
[16] *Birner, H.; Greiwe, K.*: Eine neue Klasse von eingefärbten Aluminiumpigmenten. Coating, November 1997, S. 432 – 435
[17] *Wefers, K.; Misra, C.*: Oxides and hydroxides of aluminium. In: Alcoa Laboratories, 1987, Technical Paper No. 19
[18] Showa, USP 5,037,475
[19] Toyo, EP 0 810 270
[20] Toyo, EP 0 755 986
[21] Silberline, EP 0 737 212
[22] Silberline, WO 96/13528
[23] Eckart, DE 195 01 307
[24] *Henglein, F.; Greiwe, K.*: Colour pigment coated aluminium flakes. 6th Nürnberg Congress 2001, Paper 25

[25] *Henglein, F.; Ahne, K.; Greiwe, K.*: Eingefärbte Aluminiumflakes – farbstark und rein. Farbe & Lack, July 2001, p. 94 – 99

[26] *Henglein, F.; Ahne, K.*: Coloured aluminium flakes for powder coatings. Powder Coating Europe, January 2002, p. 195 – 1999

[27] *Kiehl, A.; Greiwe, K.*: Encapsulated aluminium pigments. Progress in Organic Coatings, December 1999, Vol. 37, p. 179 – 183

[28] Eckart, DE 198 20 112

[29] *Schröder, J.*: Morphologie organischer Pigmente unter besonderer Berücksichtigung der Kupferphtalocyanine. Farbe & Lack, January 1984, p. 6 – 18

[30] *Schröder, J.*: Messmethoden zur Beurteilung der Feinverteilung von Pigmenten. Farbe & Lack, January 1988, p. 9 – 14

Guiding formulation
Medium solid effect paint, gold Rec. No. 35-01001 ECKART

1.	effect pigment	STAPA Aloxal PM ...	7.0	www.eckart.net
2.		Butyl acetate	5.5	
3.		Butanol	1.5	
4.	wetting agent	Disperbyk 162	0.4	www.byk-chemie.com
5.	anti-settling agent	Ceratix 8461	17.0	www.byk-cera.demon.nl
6.	polyester binder	Setal 90173 SS 50	38.0	www.resins.akzonobel.com
7.	melamine cross-linker	Setamine US 138 BB 70	10.5	www.resins.akzonobel.com
8.		Butyl acetate	3.0	
9.	orientation additive	CAB 381-0.5 (25 % in butyl acetate)	17.0	www.eastman.com
10.	flow additive	Byk 320	0.1	www.byk-chemie.com
			100.0	

Manufacturing procedure:

Add pos. 4 to pos. 2 and pos. 3.
Add solution to pos. 1 and prepare homogenous pigment slurry.

Attack pos. 5 approx. 10 minutes with dissolver at 5 m/s before use.

Mix pos. 6 to pos. 10 one after the other while stirring and add also the attacked pos. 5.

Finally mix this binder solution with pigment slurry.

Thinner: butyl acetate
Spraying viscosity: 17 sec DIN cup 4 mm
Drying: 30 min 120 °C

Guiding formulation
Metallic effect paint, red green Rec. No. 35-03002 ECKART

Stapa Aloxal PM 3010	3.0	www.eckart.net
Xylene	3.0	
Disperbyk 162	0.3	www.byk-chemie.com
predisperse		
Vialkyd AN 951, 70% in high-flash naphtha	15.0	www.ucb-group.com
CAB 381-0.1, 18% in butyl acetate 85	55.0	www.eastman.com
Resamin HF 480	2.5	www.ucb-group.com
Maprenal MF 590, 55% in isobutanol/xylene	4.7	www.ucb-group.com
Butyl glycol	5.0	
Sunfast-Red 202 228-6725	2.9	www.sunpigments.com
Phthalocyaningreen PG 7, 10%ige pigment paste	2.0	
Xylene	1.4	
Dowanol PMA	1.0	www.dow.com
Butyl acetate 85	4.2	
	100.0	

Spraying viscosity: 20 sec. DIN cup 4 mm
Thinner: butyl acetate 85/xylene 1:1
Spraying pressure: 4 bar
Drying: 30 min. at 120 °C

Figure 1: Liquid coating containing various types of organic colour pigments combined with aluminium pigments (substrate: steel panels) ref: Chapter IV 1.1.1

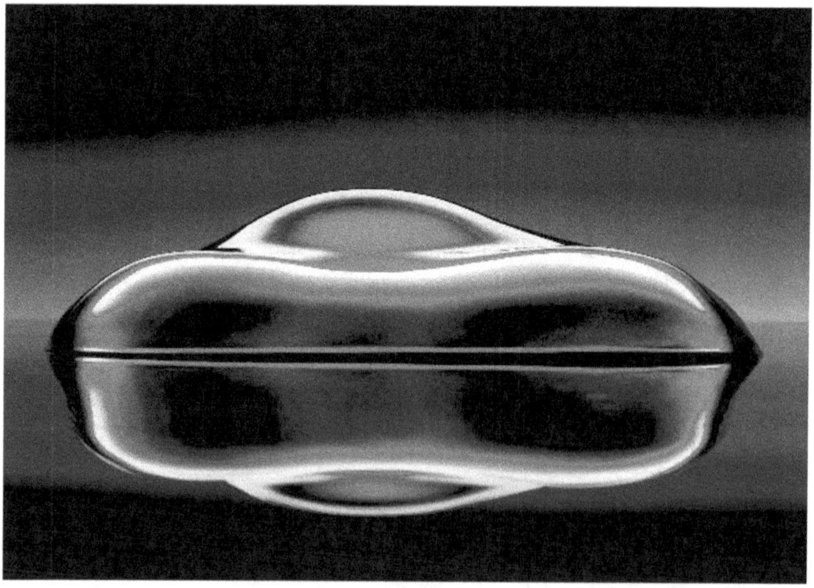

Figure 2: Liquid automotive coating containing organic colour pigments combined with aluminium pigments, ref: Chapter IV 1.2.1

Figure 3: *Liquid single layer plastic coating containing coated goldbronze pigment*
ref: Chapter IV 1.2.2

Figure 4: *Powder coat containing aluminium pigments for alloy wheel*
ref: Chapter IV 1.1.3

Figure 5A: Mass colouration of plastic item with aluminium flakes and organic colour pigments, ref: Chapter IV 3

Figure 5B: Mass colouration of plastic item with goildbronze flakes and organic colour pigments, ref: Chapter IV 3

Figure 6A and 6B:
Aluminium flakes being used as expanding agent for production of autoclaved aerated concrete, ref: Chapter IV 4

Figure 7: *Offset label print using PVD pigments ref: Chapter III 1 and Chapter IV 2.1*

Figure 8: *Nail polish with PVD pigments ref: Chapter III 1 and Chapter IV 6*

Part IV Applications

Metallic effect pigments can clearly be characterised on the basis of their manufacturing process and technical parameters such as particle size and particle size distribution, brightness, etc. A detailed description was provided in previous chapters. An exact characterisation of metallic effect pigments enables them to be classified into individual types and product lines. This classification is also essential for quality control, which guarantees a high degree of product consistency according the agreed standards.

Besides characterisation, also the area of application of metallic effect pigments is of major importance since they perform differently in each application. Thus not each individual pigment is suitable for every application. The metal effect pigments must fulfil certain base requirements to meet a specific optical and technical property profile of the application. The following chapter describes the properties of metallic effect pigments in various areas of application.

1 Coatings systems and their applications

1.1 Coatings systems

1.1.1 Solvent-based coatings

Current aspects in formulating metallic effect pigments in solvent-based coatings are discussed in Chapter 2 of Part II. Further important formulation criteria for are given here.

Slurry preparation with metallic pigments

Dispersion of pigments is particularly important and therefore discussed in detail here. Dispersion is the procedure where individual pigment particles are wetted with solvent so that they exist in mono-dispersed state (pigment slurry). The slurry is prepared using a suitable solvent in the ratio between 1:1 and 1:2 (pigment/solvent). Practical experience has shown that the 1:2 ratio is particularly of advantage when fine (average particle size less than 12 µm) or coated pigments with a large specific surface area are being used. Sometimes wetting additives must also be added during dispersion. It can also be helpful in some cases to allow a soaking time of the pigment/solvent blend of up to 10 hours – before active dispersion is started.

Polar solvents are favourable as they have a high wetting potential for metal flakes. They are only suitable for non-leafing pigments as wetting is not desired with leafing types. Leafing grades would lose their effect and thus should be made into a paste with non-polar solvents. The dissolver disc must not have any sharp corners or edges otherwise the ductile pigment can be easily damaged. A high speed mixer rather than dispersing equipment is preferable. The stirrer should be centrally positioned in the stirring container and the diameter of the disc should be at least half that of the container. Careful conversion to pastes is particularly important with gold bronze grades, which are often supplied as powders.

Pigment content

For coatings with leafing aluminium pigments, pigment content between 8% and 20% is used depending on the fineness. The pigment content for non-leafing aluminium pigments in full-shade metallics ranges from 4% for fine grades (d_{50} ca. 12 µm) to 15% for coarse grades (d_{50} ca. 40 µm). Exceptions to this are wood coatings and thermally stable coatings which can have up to 35% pigmentation. In polychromatic mixtures, significantly smaller amounts of metallic effect pigments are used. Depending on the actual shade, the pigment content ranges from 0.5% for dark tones to 3% for bright colours. Compared with aluminium flakes, gold bronze pigments have a lower opacity and copper and brass pigments are generally used with twice the pigment loading of aluminium pigments of the same particle size.

Pigment orientation

For the application of metallic effect pigments in coatings it is important to achieve optimal orientation. Various factors have an influence on this. The most important factors include:

- dilution with solvent mixtures containing both rapid and slow evaporating solvents. (The mixing ratio depends on the ambient temperature and method of application – as an approximate figure a 1:1 mixture of xylene and butyl acetate can be recommended)
- avoidance of fillers
- use of orientation promoters such as waxes and/or cellulose acetate butyrate

Pigment orientation and solids

Good pigment orientation is always obtained in paints with low solids. This can lead to film shrinkage and thus optimal pigment orientation parallel to the substrate surface (Figure IV-1.1.1.1). From this point of view, traditional low-solid systems are preferred over high-solid systems. However, low solvent emissions

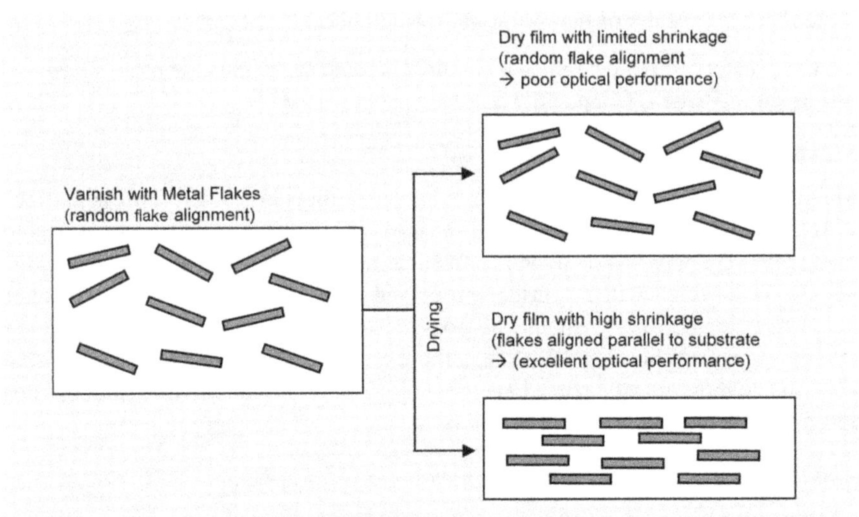

Figure IV-1.1.1.1: Orientation of metallic flakes in a coating during drying process

and the consequently lower environmental impact argue in favour of high-solid coatings.
Low-solid coatings typically have a pigment/binder ratio of about 1:5 (solid/solid) whereas for high-solid coatings the ratio is in the range 1:10 to 1:15.

Stability of metallic effect pigments

As already indicated, aluminium is an amphoteric metal which is attacked by both acids and alkalis. In many coatings, the thin oxide layer on the metal surface which is formed during grinding is sufficient to protect the aluminium pigment in the coating from the attack of aggressive agents in the environment. This is especially true for non-leafing pigments which are enclosed in a binder matrix. Basically, the purer an aluminium pigment, the better its chemical resistance and aluminium pigments are therefore normally offered with aluminium purity greater than 99.7% and in a highly pure, chemically resistant form. Acid resistant types are suitable for use in formulations with aggressive components, for example raw materials with a high acid number. Any contamination with water (for example from cleaning procedures or from impure solvents) leads to liberation of gas. In case of doubt, molecular sieves or other inhibitors should be used. The position of gold bronzes (that is copper and copper-zinc alloys) in the electrochemical series means that they are chemically even more susceptible. The mechanism of de-zincing has been described in Part II, Chapter 2.
It is therefore essential when using gold bronzes to use low-reactivity raw materials with low numbers of functional groups. It is also important to ensure water-free formulations. Green coloration and, in advanced stages, full gelling of the system are the results of inappropriate formulation.

If even greater pigment stability is required, coated types must be used. These are discussed in Part IV, Chapter 1.1.2.

Sedimentation

Metallic effect pigments settle when metallic coatings are stored. This is particularly true for specific heavy gold bronze pigments. Consequently, suitable thixotropic agents must be added to the coating especially if pigment concentrates are used which will be further processed in a mixer. Sedimentation cannot always be avoided but it should be possible to be re-stirred in a manner which avoids excessive shearing of the metallic effect pigment. Before further processing, the coating must be stirred and homogenised to avoid colour variations caused by uneven, inconsistent pigmentation.

Application of coatings containing metallic effect pigments

Coatings containing metallic effect pigments tend to flow down when applied to vertical surfaces. This sagging causes marked greying of the coating and often occurs with airless spraying in cases where high film thicknesses are applied in a single pass (cf. p. 143 et seq.). It is therefore essential that rheology is adjusted accordingly.

Clouding

Orientation of the pigments especially on large surfaces must be homogeneous. Otherwise, differences in pigment orientation can lead to differences in brightness which are visible as cloudiness. This frequently occurs with water-based coatings and must be prevented by properly designed formulations. Cornflake aluminium pigments are little susceptible to clouding as the flakes become entangled with each other via their edges and corners and so are self-fixing. Because of their rounded particle morphology, silver dollars do not have this possibility and therefore have a greater tendency to clouding.

Higher pigmentation and coarser pigments further reduce the tendency to clouding. During application, the passes should overlap as far as possible; the coating thickness should be uniform and the spraying pressure low (guide value 4 bar), in order to produce as homogeneous an orientation of the metallic effect pigment as possible. Quicker film shrinkage – by using highly volatile solvent(s) – and a low solid content further contribute to a reduction of clouding.

Use of PVD pigments

The use of PVD pigments (cf. Part III, Chapter 1) requires use of special coatings. This is the only way of ensuring good orientation of the pigment particles and reproduction of the chrome effect. Coatings formulations used for normal aluminium or gold bronze flakes will result in an unsatisfying effect. When using PVD pigments it is important to have the lowest possible solids content, ideally between 5% and 8%. Pigmentation level of the PVD pigment lies between 3% and 6%. With higher amounts, the PVD flakes lie disordered over each other which make it difficult to produce a mirror-like effect. Orientation promoters such as cellulose acetate butyrate and/or waxes should be included in the formulation to optimise the plane parallel orientation of the PVD particles. Furthermore, with the low solid content, every imperfection in the substrate surface diminishes the chrome effect. The thin (2 to 4 µm) film layers cannot level out imperfections and the structure of the substrate is then noticeable. Therefore a high-gloss, good-flow primer should always be applied before the PVD coating. Dust particles on the substrate also reduce the effect.

Experience shows that coatings with low amounts of binder have only limited resistance to moisture, etc. Scratch resistance is limited even when two-component lacquers are used and therefore over-coating is advised if excellent scratch resistance is required. The solvent used in the clear coat can initiate dissolution of the coating containing the PVD pigment, disorienting the PVD pigments and thus reducing the optical effect. For these reasons clear coats with non-polar solvents that are with low re-dissolving effect should be used. Overcoating with a water-based or powder coating, which inherently have only a low tendency to initiate dissolution, is ideal.

Guiding formulation **Base coat for polyamide substrates** **air drying**	Rec. No. 50-09018 A	ECKART
Aluminium paste ...	6.0	www.eckart.net
Butyl acetate 85	8.0	
predisperse		
Synthalat A 333, 40% supply form	41.3	www.synthopol.com
Aerosil 380	0.3	www.degussa.com
Resin: Haftharz LTW, 60% supply form	4.1	www.degussa.com
Xylene	6.7	
dissolve		
Cerafak 100, 10% supply form	7.9	www.byk-cera.nl
CAB 551-0.01, 20% in butyl acetate 98	17.1	www.eastman.com
Butyl acetate 85	8.6	
	100.0	

Viscosity: app. 21 sec, DIN cup 4 mm /20 °C

Adjust to spraying viscosity of 16 sec with the following mixture:

Butyl acetate 98	25.0
Ethyl acetate	20.0
Xylene	30.0
Ethoxy propyl acetate	25.0
	100.0

1.1.2 *Water-based coatings*

Water-based coatings offer an environmentally-friendly alternative to solvent-based coatings. Because of their relatively low solid content there is a significant volume contraction of the pigmented binder layer when water evaporates on drying or stoving. This leads to plane-parallel orientation of the pigment particles resulting in a very good metallic effect with a quality comparable to that of conventional, solvent-based systems [1].

Aluminium pigments react with water liberating hydrogen according to the following equation:

$$2\ Al + 6\ H_2O \rightarrow 2\ Al(OH)_3 + 3\ H_2\uparrow$$

When using aluminium pigments in water-based coatings it is therefore important to prevent this chemical reaction from occurring. Making the aluminium surface water repellent by use of lubricants such as oleic or stearic acids in the already described ball mill process is not sufficient. Because of the large specific surface area of aluminium pigments (1 to 10 m^2/g, BET method [2, 3]) the reaction can be very vigorous. Even in case this reaction occurs slowly, there is loss of metallic gloss and storage stability is affected (agglomeration). To prevent the undesirable gas formation, aluminium pigments in water-based coatings need efficient corrosion protection not only in the liquid paint but also for the complete lifetime of a coating [4].

In practice there are three ways of achieving corrosion protection. One possibility is to use corrosion inhibitors with high affinity for the aluminium surface and thus, within limits, form the protective layer. Typically organo-phosphorous compounds are in use (see below). Secondly, chemical reactions can be used to produce so-called conversion layers. These result out of a reaction between the aluminium surface and the corrosion inhibitor. From a chemical point of view, formation of conversion layers marks the borderline between inhibition and encapsulation. Such layers are certainly much thinner than true encapsulation but have a comparable barrier effect. The third stabilising principle is encapsulation of the aluminium pigment in a chemically inert and essentially transparent layer. Such homogeneous protective layers are usually produced by sol/gel processes (see below).

Of importance in coatings practice are the applicational properties of the effect pigment and these will therefore be considered in detail. In a second section the properties decisive for satisfactory application in water-based coatings will be described using as an example special, newly-developed, SiO_2-coated aluminium pigments. The properties are:

- wetting in the binder matrix
- visual appearance
- gassing stability
- resistance to condensation
- resistance to shear
- weather resistance
- storage stability

The examples described in the Application properties Section can be transferred to all aluminium pigments for water-based coatings. Formulation advice makes up the final part of this chapter.

Additive process and organo-phosphorous stabilisation

It is well known that compounds of phosphor have a great affinity to aluminium. Esters of phosphoric acid and substituted phosphoric acid derivatives have proved particularly suitable for inhibiting corrosion of aluminium [5 to 7]. If necessary the pH of these chemicals can be adjusted by addition of suitable neutralising agents. In practice, these organo-phosphorous compounds are blended or "dried onto" the aluminium pigment by a mixing process or a combined mixing and drying process. By adsorption onto the pigment surface they form a relatively thick protective layer. Compared to untreated aluminium pigments this results in a significant and permanent reduction of hydrogen formation in mild water-based paints. As additives are applied onto the pigment surface, this type of stabilisation is known as the "Additive process".

Organo-phosphorous protective layers make it possible to supply aluminium pigments as water-based pastes which can be formulated without any organic solvent in delivery form. This is particularly important when there is incompatibility of the solvent with the binder or for some reason organic solvents are not tolerable. Examples of completely solvent-free water-based pastes are aluminium pigments of the STAPA® Hydroxal W range (Eckart). Aluminium pigments of the STAPA® Hydrolac W, Hydrolac BG or PM ranges (Eckart) contain solvent/water mixtures and are manufactured without intermediate drying-step.

As the adsorption layers are a result of an adsorption/desorption equilibrium, adsorption layers on aluminium pigment surfaces can never completely prevent liberation of hydrogen. Moreover organo-phosphorous layers lose their inhibiting effect at high pH. This means that encapsulated aluminium pigments are necessary for particularly aggressive water-based industrial or automotive finishes. However, organo-phosphorous inhibition of aluminium pigments offers sufficient stabilisation for many water-based applications and is recommended for water-based coatings with pH between 7 and 8, such as industrial coatings or mild automotive finishes and water-based printing inks. The suitability of pigments stabilised in this manner should always be confirmed by a gassing test (cf. Part II, Chapter. 3: Accelerated gassing stability test).

A further possibility for stabilising aluminium pigments is the addition of anti-corrosive pigments to coatings systems or pigment pastes. These are mainly mixtures of phosphate silicates which are soluble only with difficulty. The protective effect has two origins (i) formation of phosphate at the reactive centres of the aluminium particles and consequently the barrier effect of a thin protective layer, and (ii) the electrochemical effect. Aluminium pigments stabilised in this way are commercially available as pigment preparations in the form of pastes, dispersions and granules

Conversion layers by chromating

A well-known technique for passivating aluminium pigments is chromating. In this aluminium pigments are treated with a solution of chromic acid [8]. This rapidly oxidises the reactive aluminium surface and forms a thick, passivating conversion layer. This is mainly a mixed oxide consisting of aluminium oxide, chromium oxide and the corresponding hydroxides. A similar method of passivation is based on using molybdic acid [9].

Chromated aluminium pigments have excellent gassing stability in water-based coatings and are even stable in very aggressive water-based formulations. If necessary they can tolerate pH values up to about 9.5. Additionally this type of treated aluminium pigment has very good resistance in condensation tests (DIN EN ISO 6270-2). The processing properties of chromated aluminium pigment pastes with common co-solvents and coatings additives are very good. The low viscosity of such semi-finished products (pigment slurries) provides easy incorporation into water-based binders.

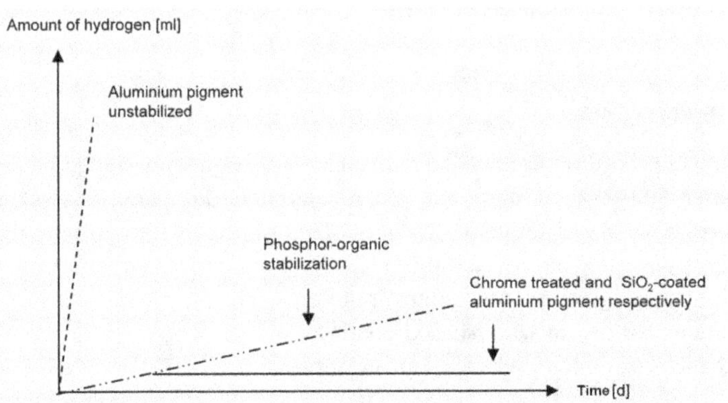

Figure IV-1.1.2.1: Gas evolution of various stabilized aluminium pigments

Figure IV-1.1.2.1 illustrates the performance of untreated, organo-phosphorous-stabilised and chromated aluminium pigments in aqueous coatings with regard to gas evolution in a special gassing test. It is evident that chromated metal flake pigments have the best gassing stability. They show no or at least extremely little gas development in water-based coating systems even after storage for many months. This is their real advantage compared to pigments with organo-phosphorous stabilisation as these show a small but continuous liberation of hydrogen in aggressive varnishes.

Encapsulation

The purpose of encapsulation is to coat the surface of the aluminium pigment completely with an inert protective layer and in this way to protect it from reaction with water. The optical and applicational properties of the pigment should be as little affected as possible by the encapsulation. This method offers the possibility of applying organic and inorganic coatings of different thicknesses (barrier effect). The chromated pigments already described form the standard against which encapsulated aluminium pigments are judged. The aim is for heavy-metal-free encapsulation with polymer or silicate layers which were until recently only successful in certain special applications. These encapsulations, which are interesting for ecological reasons, are considered below.

Polymer Coating

Polymer encapsulation involves first coating the aluminium pigment with a primer layer and then coating it by radical polymerisation of multi-functional acrylate monomers [10]. Such polyacrylate layers however have a high porosity which makes these aluminium pigments suitable for water-based coatings only to a very limited degree. In powder coatings however, polymer coated aluminium pigments show good corrosion resistance in weathering tests. An example of these is the STANDART® PCA range (Eckart).

Modified SiO_2 coating

Coating process

SiO_2-coated aluminium pigments have been known for a long time [11]. However conventional SiO_2-coated pigments are not suitable for water-based coatings. Thanks to a new sol-gel process, silicate-coated pigments can now be used in a wide variety of water-based coatings [12]. The excellent properties mean that they are in no way inferior to chromated pigments. The profile required for aluminium pigments for use in water-based coatings will now be discussed extensively using these new developments as an example.

The process of SiO_2-encapsulation of aluminium pigments starts with dispersing the uncoated pigment in an alcoholic solvent. All available aluminium pigments can be used for this. The next step involves adding tetraethyl silicate (TEOS), water and a basic catalyst to the stirred pigment suspension. Hydrolysis of the silicic acid ester formally leads by a multi-step process to the formation of silicic acid [$Si(OH)_4$], which immediately polycondenses, releasing water. In this polycondensation finely dispersed SiO_2 gel particles are formed which deposit very evenly on the surface of the pigment, under suitable reaction conditions.

$$Si(OR)_4 + 4\,H_2O \xrightarrow{\text{Base}} [Si(OH)_4] + 4\,ROH \quad \text{(Hydrolysis)}$$

$$[Si(OH)_4] \xrightarrow{\text{Base}} SiO_2 + 2\,H_2O \quad \text{(Condensation)}$$

Many technological properties of metallic effect pigments depend on their surface chemistry. Sol/gel chemistry has the advantage that freshly deposited silicate layers can be modified in a controlled manner. Bifunctional reagents, as for example organo-silanes, are suitable for such surface modification.

Industrial process

The process is carried out in a reactor which is heated and stirred (Figure IV-1.1.2.2). Surface modification can follow conversion of the silicic acid ester to SiO_2. The reactor is then emptied and if necessary the pigment suspension separated from agglomerates. Excess solvent is removed using a filter press. Finally

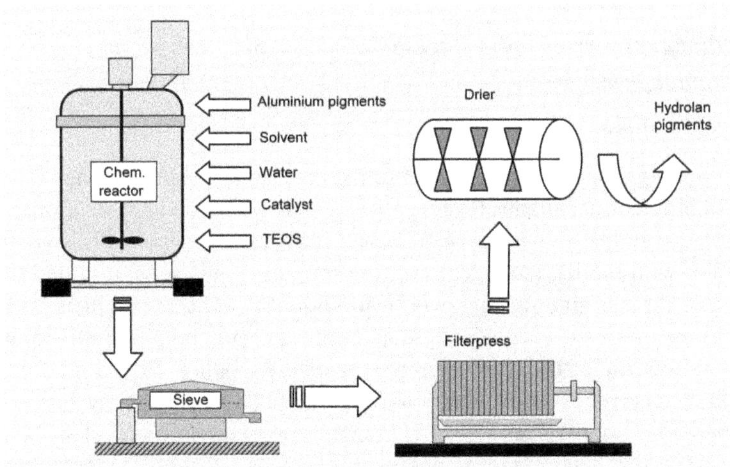

Figure IV-1.1.2.2: Sol-gel coating of aluminium pigments

the pigment is dried and converted into a paste with the required solvent. The complete production can be carried out as a batch process [13].

Scanning electron microscope images (SEM) show that the aluminium flakes are fully coated by a thin, homogenous silicate layer. Precipitation of fine SiO_2-particles is not apparent (Figure IV-1.1.2.3). The specific surface area of the aluminium pigment is practically unchanged by the coating process (determined by BET [3]). From this it can be deduced that the thin silicate layer is very dense and has only few pores.

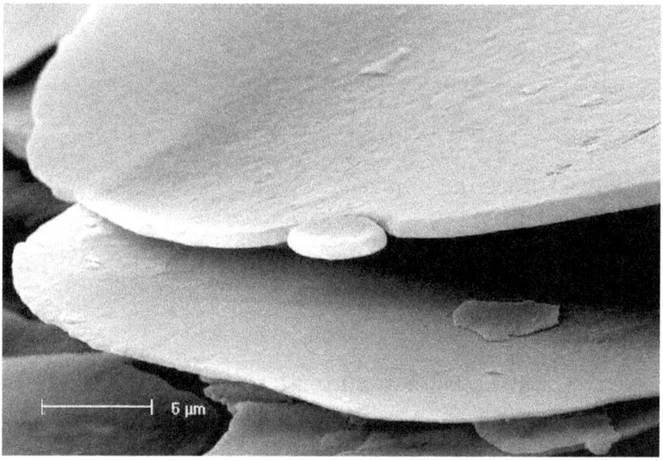

Figure IV-1.1.2.3 : SEM view of coated aluminium flake

Applicational processes: advantages of modified SiO_2-coating in applicational processes

Controlled wetting

Aluminium pigments can be adjusted to the coating system by surface modification with bifunctional silanes (Figure IV-1.1.2.4). When the first functional group has reacted with the SiO_2-surface, a second functional group is still available for interaction with the basecoat. The wetting properties, dispersability and the orientation of the aluminium pigments in the liquid coating are influenced by the chemical nature of the secondary group. When stoving the paint, this group can react with the binder or other components of the coating. Bonds of a physical (Van-der-Waals forces, hydrogen bonds) or chemical nature can be formed. The latter markedly improve resistance to condensation and promote adhesion of the complete coating. The pigment effectively becomes a firmly bonded component of the coating system [14].

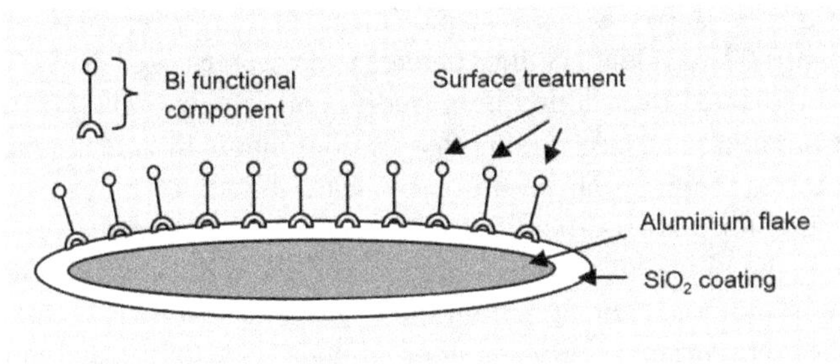

Figure IV-1.1.2.4: Aluminium pigment with SiO_2-coating and additional surface treatment

Modification of the pigment surface with non-polar groups such as alkyl groups results in a relatively poor wetting of the pigment in water-based formulations. The visible result is a so-called semi-leafing effect with parallel orientation of the aluminium pigments relatively near to the surface of the coating layer. This leads to high brilliance but under some circumstances to poorer adhesion of the coating layer.

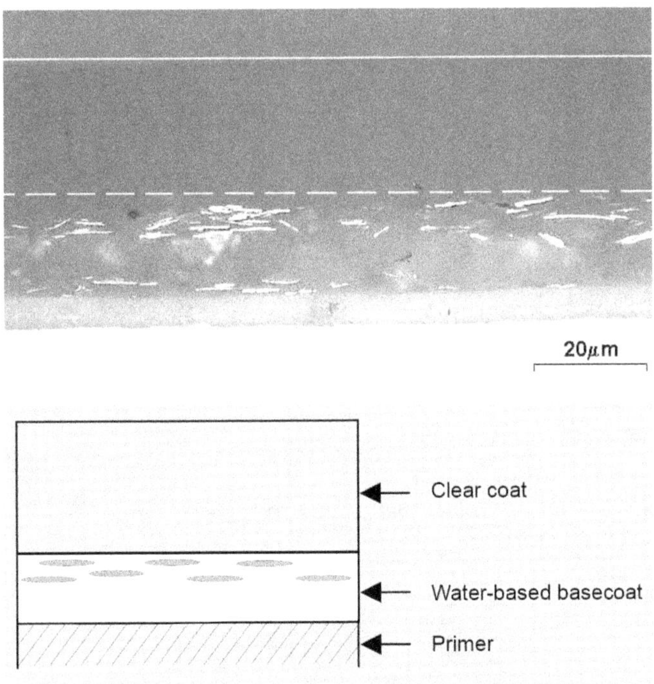

Figure IV-1.1.2.5: Cross section through a coated aluminium pigment of poor wettability (semi-leafing)

Figure IV-1.1.2.5: shows a microscopic image of the cross section of such a metallic coating. The boundary-line between the water-based basecoat and the clear coat is indicated by a dotted line. Additionally the boundary of the varnish and the surrounding matrix is marked by the continuous line. If the pigment surface is modified more polar, wetting by water-based coatings is improved. Suitable groups for this modification are acrylate, amine, epoxy and ureido. The orientation of the pigments in the coating is still good but the particles are now dispersed more homogeneously within the coating layer. Often such a modification is the best compromise between good pigment orientation and good adhesion properties (Figure IV-1.1.2.6).

If the pigment surface is very polar, as is the case with pure SiO_2, parallel orientation of the pigment will be impaired. In this case the aluminium pigments have a tendency to settle and agglomerate. As a consequence, too good a wetting of special-effect pigments often leads to loss of brilliance and reduced opacity

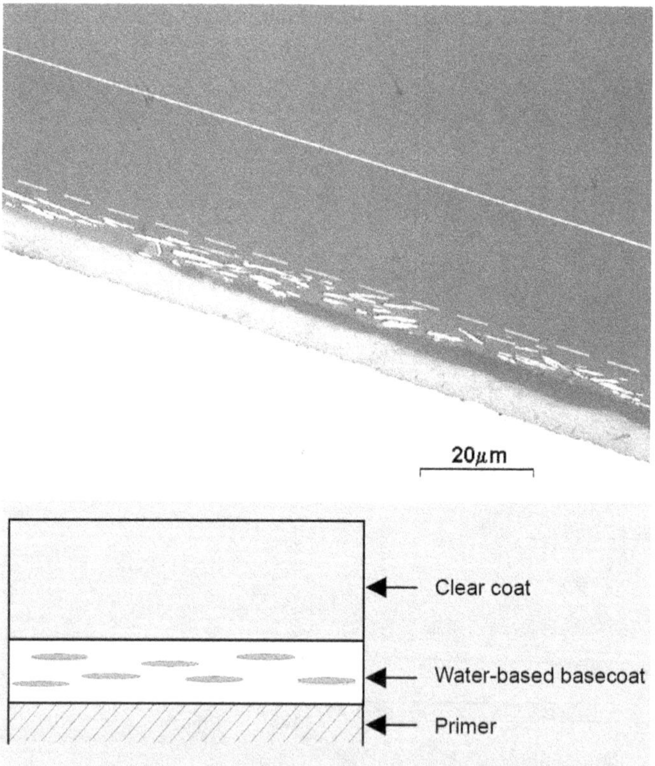

Figure IV-1.1.2.6: Cross section through a coated aluminium pigment with excellent wettability

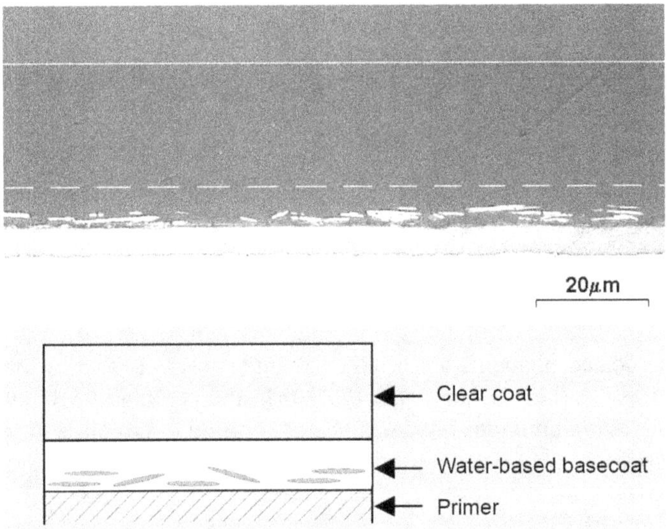

Figure IV-1.1.2.7: Cross section through a coating with settled and agglomerated pigments

(Figure IV-1.1.2.7). Naturally the wetting and dispersing properties of the aluminium pigments strongly affect the rheology of a water-based coating [15]. A freshly deposited SiO_2-layer generally has numerous silanol groups which can produce a strong increase in viscosity. Controlled substitution of these OH-groups allows the rheological properties of the pigment to be optimally adjusted to the demands of the coating manufacturer.

Brilliance

The metallic effect depends on the size and shape of the flakes and the wetting of the pigment in a given coating system. Using the sol/gel process described above followed by a modification, it is possible to manufacture pigments with Taylor made wetting properties. Every passivation or encapsulation of aluminium pigments however affects the optical properties so that the same conventional aluminium pigment can have different optical effects depending on the type of stabilisation. Table IV-1.1.2.1 compares the colour measurement results of brightness L* of SiO_2-coated aluminium pigments (STAPA® Hydrolan) and that of the same pigments, chromate stabilised (STAPA® Hydrolux). The brightness difference ΔL* is to be understood as difference (L* [SiO_2-coated] – L* [chromate passivated]). Positive ΔL*-values denote greater brightness of STAPA® Hydrolan, and vice-versa.

Pigment STAPA®	Particle size distribution (Cilas)			Difference in brightness ΔL*		
	d_{10}	d_{50}	d_{90}	25°	45°	75°
Hydrolan 2192	8	15	26	2.39	0.16	-0.95
Hydrolan 2156	10	17	28	2.38	−1.21	-1.56
Hydrolan 501	8	21	41	4.49	1.48	-1.58
Hydrolan 167	11	25	45	3.62	−1.29	-3.60
Hydrolan 214	17	34	57	3.54	−1.00	-2.23

Table IV-1.1.2.1: Brightness difference ΔL of chromate and SiO_2-coated aluminium pigments*

All silicate coated pigments are brighter at angles close to that of the incident light (25°) and darker on flop angle (75°) than the chromated reference types. That means that aluminium pigments coated with SiO_2 exhibit a stronger flop, obviously because they are better oriented and give a more metallic appearance.

Circulation line stability

In paint application lines of the automotive industry, water-based coatings are continuously pumped around and are kept circulating. High shear forces can damage the surfaces of the pigment particles, especially if gear pumps are used. This strong mechanical shear stress is usually simulated by the Waring-Blendor Test in which a sharp edge stirrer configuration is used with very high speed to agitate the basecoat (15 min. at 20,000 rpm.). This test is in use for high-solid and water-based coatings. The test results are assessed by comparing the applications of paint before and after application of shear.

Because of their sandwich structure, SiO_2-coated aluminium pigments are well stabilised against mechanical impacts. Figure IV-1.1.2.8 illustrates that silicate coated aluminium pigments (STAPA® Hydrolan IL 8154) have far greater shear stability than uncoated (STAPA® Metallux R 8754) or chromate passivated pigments (STAPA® Hydrolux PM Reflexal 8154). They have the same mechanical stability as special shear-stable aluminium pigments (STAPA® NDF 160; so-called. Tufflakes – cf. p. 34 f), without their disadvantage of reduced hiding power.

Shear stability of pigments is divided into three categories depending on their brightness difference ΔL* at 20° in the CIELAB-System:

- ΔL* < 5: non degrading (stable to shear)
- ΔL* = 5 – 10: semi-degrading (moderately stable to shear)
- ΔL* > 10: degrading (not stable to shear)

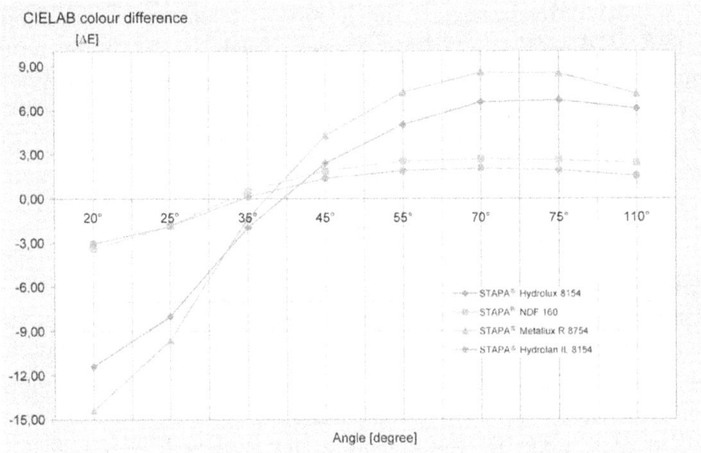

Figure IV-1.1.2.8: Comparison of mechanical stability of aluminium pigments

Further important applicational characteristics

Gassing stability

Aluminium pigments have their best gassing stability at pH 7 and the stability drops considerably in more acid or more basic media. Most water-based coatings are alkaline as the binders contain carboxyl groups which have been neutralised with an excess of amines. Soluble resins are particularly difficult for aluminium pigments as they often have a pH greater than 8. Sol/gel-coated pigments are particularly suitable for use in such corrosive media and show excellent stability in the 40 °C gassing test. In fact the only things they are comparable with are chromated pigment grades. No dependence is found on the nature of the binder and its pH. Even with storage for a 3 months period the total gas evolution is typically less than 5 ml of hydrogen. This corresponds to less than 1/2000 of the total weight of aluminium pigment in 300 ml of water-based coating (with 3% pigmentation). Similar as with chromated aluminium pigments, a very small amount of initial gassing may occur during the first few days. After that there is no further gassing. The apparatus for the 40 °C gassing tests is shown in Figure II-3.2 (Part II, Chapter 3.).

Optimised condensation resistance in automotive applications

The necessary adhesion tests required by these grades are carried out on a standard commercial automotive OEM-system, consisting of five coating layers: phosphatisation (1), cataphoretic coating (2), filler (3), metal pigmented water-based coating (4) and clear varnish (5). The joint effect of all these layers determines the durability properties of the coating.

The standard test for resistance to condensation of a coating with aluminium pigments is DIN 50017-KK (ISO 6270). The test panels are exposed to water-saturated atmosphere at 40 °C for 240 hours. After this adhesion is immediately evaluated using cross-cut test in accordance to DIN EN ISO 2409. Table IV-1.1.2.2 summarises the effects of surface modification on condensation resistance and shows that the wetting properties of the SiO_2-layer can be adjusted by specific selection of surface modifiers.

If SiO_2-coated pigments are used without any surface modification, the condensation resistance is insufficient (GTC = 4 to 5). The SiO_2 is strongly hydrophilic and water penetrates easily between the pigment surface and the binder matrix leading to delamination of the upper coating layers. With a surface modification which allows only a small amount of wetting such as alkyl groups, the condensation performance is significantly improved but still insufficient. The best results are obtained with moderate wetting properties achieved by modification with amino or acrylate groups.

Funktionalität	Cross-cut test results		
	1 min	1 hr	24 hrs
pure SiO_2	Gtc 5	Gtc 4	Gtc 4
alkyl	Gtc 2	Gtc 2	Gtc 1
amino	Gtc 0	Gtc 0	Gtc 0
acrylic	Gtc 0	Gtc 0	Gtc 0

Table IV-1.1.2.2: Adhesion after condensation test

In automotive finishes, the adhesion of the complete multi-layer coating system is tested using impact tests. This is done by impact of little stones or steel balls with defined energy (stone-chipping test) on the surface. Consequently the degree of damage is evaluated.

Weather resistance

In automotive industry, testing of weather resistance of a coating (in the so-called Florida test) is of major importance. In this outdoor durability test, a series of pigments are tested in OEM finishes by exposure to extreme climatic conditions. Tests using various aluminium pigments show that durability and UV stability of silicate-coated (and equally chrome and additive after-treated) aluminium pigments in commercial OEM- and refinish systems are excellent. After 2 years of outdoor weathering, brightness, colour strength and colour saturation of the test specimen were practically unchanged ($\Delta E < 1$ at all angles of measurement). Non-exposed panels were used as references (dark-standards) in any case.

Storage stability

Storage stability of solvent-containing pigment pastes based on SiO_2-coated aluminium pigments is also excellent. It is at least 1 year (at 25 °C) and is therefore much better than that of pastes of uncoated aluminium pigments. Even at 50 °C no significant optical changes are found after two months storage.

Stabilised aluminium pigment grades

Table IV-1.1.2.3 gives an overview of the different structures and uses of stabilised aluminium pigments in water-based coatings.

Pigment type	Pigment orientation	Applications	Special applications
Aluminium pigments with corrosion inhibitors, e.g. STAPA® Hydrolac, STAPA® Hydroxal	leafing and non-leafing	"mild" automotive OEM basecoats, general industrial coatings	–
Encapsulated aluminium pigments, e.g. STAPA® Hydrolan	only non-leafing	automotive refinish systems, "aggressive" automotive OEM-systems, coatings for plastics	electrically isolating coatings (solvent based) for plastic parts (computer housing, TV-sets), e.g. STAPA® Hydrolan
Aluminium pigments with conversion layer stabilisation, e.g. STAPA® Hydrolux	only non-leafing	Automotive refinish systems, "aggressive" automotive OEM-systems, coatings for plastics	–

Table IV-1.1.2.3: Stabilised aluminium pigment grades

Formulation advice

Leafing pigments

Stabilisation of aluminium pigments with organo-phosphorus additives (see above) does not change the leafing- or non-leafing properties. With leafing types of the STAPA® Hydrolac or STAPA® Hydroxal range however, careful choice of organic solvent must ensure that the pigment flakes will not be wetted on storage. Water is ideal as a solvent for forming a paste. Because of its high surface tension, only a very weak or even no wetting agent should be added to assist the dispersion of the aluminium pigment. Basically, aluminium pigments which contain white spirit in delivery form (STAPA® Hydrolac) have better leafing characteristics. The contrast between the polar solvent water from the formulation and the non-polar white spirit from the pigment paste reinforces the leafing behaviour of the metal flake pigment. Because of their favourable price/performance ratio, coarse products of the STAPA® Hydrolac range are typically used for roof coatings. Fine grades are suitable to formulate reflective coatings and chrome effects.

Non-leafing grades obtained by the additive process

Non-leafing grades are mainly used in industrial and plastics coatings. Again the reason is the favourable price/performance ratio. Aluminium pigments stabilised by an additive process are also used in deliberately formulated OEM-coatings. For dispersing these grades, pigment slurry is prepared in a 1:1 mixture of water and a suitable co-solvent. Under some circumstances, suitable surfactants are recommended as wetting-agents. However before-hand tests have to be carried out to ensure that these wetting agents do not remove the additive stabilisation from the metal pigment surface, which may result in gas formation during storage. As the adsorption layers on the metal flake pigment are in chemical equilibrium, too strong a wetting agent can remove them from the surface.

Side effects of additive-coated aluminium pigments

In paint formulations, all additives have both good and bad effects. This can also be seen with metal flake pigments stabilised against gas development by additives. In certain applications drying behaviour, water resistance and/or intercoat adhesion can be influenced negatively. To minimise such side effects, the amount of stabilisation additive can be halved or even further reduced. Resistance to gas development in every single binder system must then again be tested.

Non-leafing grades after encapsulation

The encapsulation of aluminium pigments already described leads eventually to loss of the leafing character. Therefore coated pigments are only available as non-leafing products. They are incorporated into the binder matrix in form of pigment slurry. Because of the somewhat larger specific surface area, an aluminium pigment to solvent ratio of 1:2 should be chosen, especially for fine grades with d_{50} below 15 µm. Suitable surfactants promote improved dispersion.

Application examples of encapsulated aluminium pigments

Main application of this product group is for automotive finishes. Especially in water-based refinish formulations, encapsulated aluminium pigments have significant advantages because of their excellent gassing stability. As their surface is chemically inert, there are no side effects such as have been described for the additive-stabilised types. In mono-coat paints for plastic substrates (for example in consumer electronics), coated aluminium pigments have the advantage of high chemical resistance.

Viscosity of water-based metallic coatings

Thixotropy is indispensable for optimal effect forming with water-based metallic coatings. Only this prevents sedimentation of metal flake pigments on storage. At the application stage, atomisation is achieved following high shear forces. Immediately after the paint droplets reach the substrate, the viscosity rises so that sagging is avoided. Sagging on vertical surfaces would impair orientation of the flakes so that even a high-quality pigment would appear grey and unattractive.

Clouding with water-based metallic coatings

In water-based coating applications on large surfaces there is a high potential of non-uniform pigment orientation causing "clouding". This can be avoided if a suitable orientation promoter is added to the paint: in practice wax emulsions are used. Cellulose acetate butyrate compounds, such as are already used with solvent-based systems, are still under development for water-based coatings. Choice of suitable application parameters can significantly reduce clouding. This is also positively influenced by reduction of the evaporation time. The quicker the solvent evaporates, the lower is the tendency for clouding.

Gassing of water-based metallic coatings

Both the stabilisation process of the metal flake pigment and the paint formulation affect the tendency to gassing in water-based coatings. Shear forces should be kept as low as possible during dispersion of the metal flake pigment (see above). Certain binders are more aggressive to aluminium flakes than others. Information can be obtained from the pigment manufacturer. A decisive parameter is the pH-value of the coating. This should be adjusted as close as possible to the neutral point, pH 7.

For neutralisation a mild base, such as triethanolamine or dimethylethanolamine, is recommended. Strong bases such as ammonia react with the aluminium metal and lead to a strong gassing tendency of the coating system. In formulating metal flake coatings combinations with transparent iron oxide show an increased tendency to gassing. An explanation of this phenomenon has been given above. The following scheme (Table IV-1.1.2.4) gives an overview of common problems and their solutions without however claiming to be fully comprehensive [16].

Gold bronze pigments for water-based coating systems

Gold bronze pigments are available as leafing and non-leafing grades for water-based coatings systems. Because of the high reaction potential of copper in water as already mentioned, they must be stabilised very effectively. The reaction potential decreases from pure copper through the natural tones to the fire-coloured tones. In all cases stabilisation is achieved with selected hydrophobic additi-

Problem	Possible reason	Suggested solution
Gassing of coating	aluminium pigment is not stable enough for this coating	use a better-stabilised type of water based coating
	pH value too high	change pH value: • pH8: STAPA® Hydrolac/Hydroxal, ROTOTVARIO® Aqua • pH9: STAPA® Hydrolan, STAPA® Hydrolux
		use suitable neutralizing agent
	dispersion too aggressive	use ROTOVARIO® Aqua
		use • STAPA® Hydrolac PG/PM • STAPA® Hydroxal BG/PM • STAPA® Hydrolux PM Reflexal • STAPA® Hydrolan IL
		use solvent and/or additives
Poor optical appearance	aluminium pigment with wide particle size distribution	pigment with narrower particle size distribution
	solids too high	lower solid
	rheology	formulation with thixotropy
	bad orientation	orientation agent
	poor dispersion	use dispersing agent
Poor resistance to condensation water	amount of dispersing agent too high	reduce or change
		use of other aluminium pigment grades
	amount of dispersing agent on the aluminium pigment surface too high	use • ROTOVARIO® Aqua • STAPA® Hydrolac/Hydroxal • STAPA® Hydrolac/Hydroxal H • STAPA® Hydrolac/Hydroxal HH • STAPA® Hydrolux • STAPA® Hydrolan (↓ Increase in stability)
	binder too hydrophilic	use more hydrophobic binder
Bad colour repro-ducibility	check dispersion	organic pigments: Use concentrate
		aluminium pigment: Improve dispersion

Table IV-1.1.2.4: Overview of common problems with metallic pigments in water based coatings and suggested solutions

ves. They are blended with the leafing gold bronzes. With non-leafing types SiO_2-coated STANDART® Resist or Dorolan® grades are used and mixed with the same additives. The SiO_2 coating on its own does not provide sufficient stabilisation. Some of the products named are commercially available as pourable and pumpable pigment concentrates.

Unlike aluminium flakes in water-based coatings, gold bronze systems do not gas but initially turn green because of de-zincing and dissolution of copper ions. As a result the complete paint mixture can gel or form a firm skin on the surface (cf p. 36 f). Particular care must be taken for choice of raw materials: they should only contain molecules with as few reactive groups as possible. The pH should be adjusted to values between 7 and 8 (if necessary) with a mild base. In addition all the same requirements as for aluminium pigments apply when using gold bronze pigments. For mass tone formulations, pigment loading is between 8 and 10% (solid) depending on the fineness of the pigment. The high specific gravity of gold-bronze pigments requires suitable stabilisation against sedimentation especially if water-based coatings are to be stored for longer periods of time.

Zinc pigments in water-based coatings systems

As a metal in aqueous media, zinc has extraordinarily high reactivity. It is therefore difficult to stabilise against attack by water. If an effective stabilising layer was applied, the electrical conductivity between pigment particles would be stopped and therefore a typical corrosion protection function would be lost. In practice, zinc flakes are incorporated into a water-free component and stored in this form. Only just before application, this component is mixed with a second water-containing component. The mixture is then used within about eight hours. The water-free component should be free of amine-hardener, as amines also react with zinc forming gas. Because of the high specific gravity of zinc, anti-settling agents must always be added to the paint formulation.

References

[1] *Dören, K.; Freitag, W.; Stoye, D.*: Wasserlacke. Umweltschonende Alternative für Beschichtungen. Köln 1992

[2] *Brock, T.; Groteklaes, M.; Mischke, P.*: European Coatings Handbook, Hannover 2000

[3] *Wedler, G.*: Lehrbuch der Physikalischen Chemie, Weinheim 1982

[4] *Karsa, D. R.; Davies, W. D.*: Waterborne Coatings and Additives. Cambridge 1995

[5] Toyo; EP 0240367

[6] *Ruf, J.*: Organischer Metallschutz, Hannover 1993

[7] *Müller, K.-P.*: Praktische Oberflächentechnik, Wiesbaden 1999

[8] BASF, EP 0 259 592

[9] Toyo, EP 0 633 297

[10] Eckart, EP 0 477 433

[11] DuPont, US Patent 2885336

[12] *Kiehl, A.; Greiwe, K.*: Encapsulated aluminium pigments. Progress in Organic Coatings, 1999, Vol. 37, pp. 179 – 183

[13] *Ignatowitz, E.*: Chemietechnik, Haan-Gruiten 1994

[14] Eckart, EP 1 084 198

[15] *Metzger, T.*: The Rheology Handbook, Hannover 2000

[16] *Wißling, P.*: State-of-the-art technology in aluminium pigments for aqueous paints. Surface Coatings International, July 1999, Vol. 82, pp. 335 – 339

Guiding formulation
Metallic basecoat, silver
water thinnable

Rec. No. 76-09012

ECKART

1.	Stapa Hydroxal / Hydrolan ...	4.34	www.eckart.net
	Stapa Hydrolux and Hydrolac not suitable		
	Additol XL 250	0.43	www.ucb-group.com
	Butyl glycol	4.70	
	n-methyl pyrrolidone	1.57	
2.	Dimethylethanolamine	0.23	
	Viscalex HV 30, 10 %ig*	4.23	
	Water demin.	4.70	
3.	Viacryl VSC 6265w, 40 WA	5.34	www.ucb-group.com
	Water demin.	15.67	www.ucb-group.com
4.	Viacryl VSC 6276w, 44 WA	22.78	
	Maprenal MF 900w/95	3.13	
	Butyl glycol	3.92	
5.	Viscalex HV 30, 10 %ig*	4.70	
6.	Dimethylethanolamine, 10% in H_2O demin.	0.22	
	Water demin.	7.83	
7.	n-Butanol	0.78	
8.	Water demin.	15.43	
		100,00	

*** Viscalex HV 30, 10 %ig:**		
Water demin.	64.5	
Dimethylethanolamine, 10% in H_2O	2,5	
Viscalex HV 30	33.0	www.cibasc.com
	100.0	

Production:

Blend part 1 in mentioned order and disperse for 30 min with a fast stirrer (not a Dissolver).

Add part 2 in the order shown and mix for 15 min.

Mix part 3 separately and quickly add with 10 min stirring.

Add part 4 in mentioned order and mix for 30 min.

Add parts 5 and 6 and mix for 10 min, then homogenize with part 7 for 5 min.

Adjust to application viscosity of 30 sec (DIN 53211, 23 °C) with part 8.

Filter through an 80 µm metal mesh.

1.1.3 Powder coatings

Since the 1960s, powder coatings have provided an interesting economic and ecological alternative to liquid paints. Their advantages lie in their relatively unproblematic application and the possibility to recycle powder coating particles which have missed the object to be coated (overspray recycling). Overspray can be recovered immediately; almost all the material is used. The good electrostatic wrap-around effect at the application process together with a high dry film thickness of 60 to 80 μm result in outstanding mechanical durability and excellent corrosion resistance of the coated objects. Powder coatings make no major demands on the substrate, requiring no cost-intensive surface preparations. In fact, even small grinding marks, surface defects and weld seams are hidden completely underneath the powder coating and levelled out optically as a result of the slight orange peel effect. Currently one major disadvantage compared to liquid paints is the high curing temperature required by powder coatings. Systems, such as glycidyl methacrylates (GMA systems), have however been introduced in the OEM automotive sector, for example, and require stoving temperatures of only about 120 °C. Their specific advantages and disadvantages will not be discussed further here.

The methods of incorporating metallic pigments into powder coatings and the subsequent performance differ significantly from those for liquid paints described in preceding chapters. The basic principles are explained following, using aluminium pigments as an example which is widely utilised in powder coatings. These principles also apply to a large degree to gold bronze pigments (copper- and copper/zinc flakes), whose special characteristics will be discussed at the end of this chapter.

Orientation characteristics of metallic effect pigments

As in liquid paints, two different groups of metallic effect pigments are used in powder coatings: leafing pigments and non-leafing pigments. If the pigment is oriented at the top of the coating surface, this is referred to as leafing. This behaviour is created typically by adding stearic acid during ball-milling of the metallic granules into flakes (cf. Part II, Chapter 2). As in liquid paints, stearic acid treated flakes are only poorly wetted by the binder. In the melt phase of the hot powder coating, de-mixing phenomena occur and the stearic acid coated metal particles float and are concentrated in the upper surface of the coating film. Because of the high viscosity of the powder coating binder of 2000 to 9000 mPa·s (at 200 °C, measured by the ICI-Plate-Cone method) complete floating of all pigment particles to the surface never occurs, unlike liquid paints. Figure IV-1.1.3.1 shows a cross section through a transparent powder coating manufactured by the Dry-Blend-Process with a pigment loading of 2.0% of a medium fine

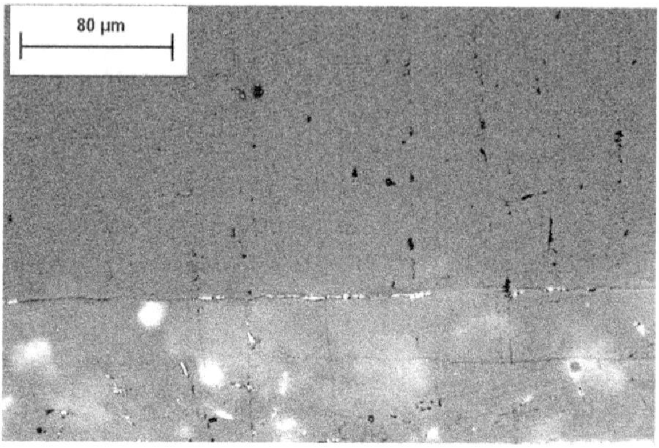

Figure IV-1.1.3.1: Dry blend with a pigment loading of 2.0% of a medium fine leafing aluminium pigment (STANDART® Aluminium Powder PC)

leafing pigment. Besides a very high proportion of leafing pigments (i.e. flakes on the surface of the coatings), a comparatively high proportion of pigment particles are embedded inside the binder matrix and are optically inactive.

Metallic effect pigments are referred to as non-leafing when the pigment particles remain embedded in the powder coating matrix. Of course, because of the high melt viscosity, a certain proportion of pigment is also found near to or on the surface. The non-leafing characteristics of metal flakes are obtained by adding unsaturated fatty acids such as oleic acid during ball-milling.

The metallic effect pigments mainly used in the powder coatings industry are post-treated in a second step after ball milling with an inorganic coating, typically SiO_2. These pigments are available from *Eckart* as STANDART® PCR-Powders and are extremely easy to handle in powder coating applications (see below). As they are encapsulated non-leafing pigments, they offer better chemical resistance than uncoated aluminium pigments.

Improved performance as a result of considerably solidified SiO_2-coating has led to a new generation of powder coating pigments. They are manufactured similar to the SiO_2-coated Hydrolan products (as described in Part IV, Chapter 1.1.2) und they are supplied from Eckart in powder form under the trade name STANDART® Sillux. Besides improved chemical resistance they also have excellent optical appearance.

Aluminium pigments coated with acrylic polymers are also used in powder coating formulations because of better handling and chemical resistance (cf. p. 88). Eckart offers these products in dry powder form (STANDART, PCA).

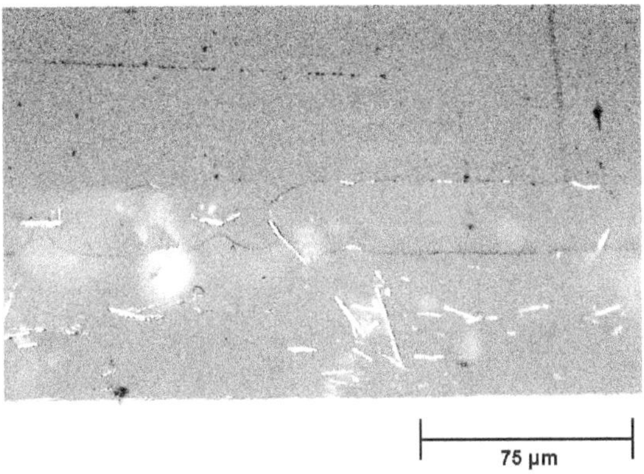

Figure IV-1.1.3.2: Dry blend with a pigment content of 2.0% of a very fine coated, non-leafing aluminium pigment (STANDART® Aluminium Powder PCR 801)

Figure IV 1.1.3.2 shows a clear coat with 2.0% pigment loading using a very fine coated, non-leafing aluminium powder, dry-blended into the powder coating (analogous to Figure IV-1.1.3.1).

Incorporating metallic pigments

Three different processes are currently used to incorporate metallic effect pigments in powder coatings. These processes have significant advantages and disadvantages.

Dry Blending

In Dry Blending, the oldest and still the most common manufacturing process, the powder coating is filled into a slowly operating mixer (e.g. a tumble mixer). The appropriate amount of metallic effect pigment is then added. Safety precautions should be taken since there is a risk of dust explosions when working with aluminium pigments (see below.). The addition of a small amount of flow agent (in powder form) is beneficial. After mixing is complete, the powder coating is usually passed through a sieve to separate out coarse material and agglomerates. In this process, powder coating and pigment particles end up lying randomly arranged next to each other.

The advantages of Dry Blending are quick and cheap production of powder coatings. The main disadvantage is that during application, separation of the powder coating and metallic effect pigment particles can easily occur and, in particular the relatively light aluminium pigments can easily be removed into the extraction system. As a result, Dry Blend products are not considered recyclable since the desired metallic effects are gradually lost. A further disadvantage is incomplete dispersion of metallic pigments and their tendency to reagglomerate in the powder gun. Incomplete dispersion leads to accumulations of pigment on the coating surface which can be seen as micro-cloudiness. (cf. Figure IV-1.1.3.3). To ensure batch-to-batch consistency of Dry Blends, production parameters, such as mixing unit, mixing time and speed, must be controlled tightly at any time, otherwise colour shifts can easily occur.

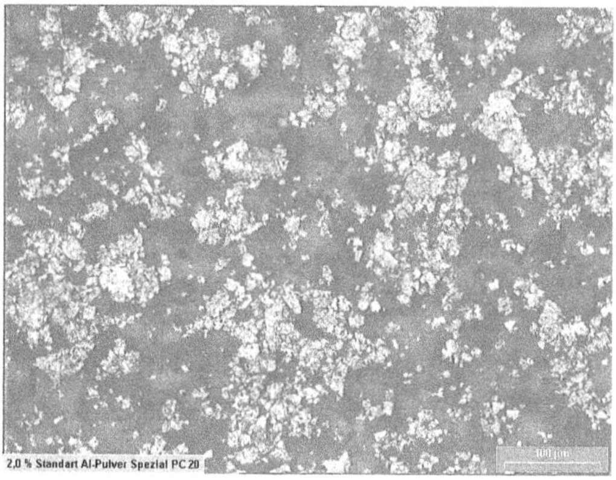

Figure IV-1.1.3.3:
Coating with cluster formation when using a Dry Blend with a pigment loading of 2.0% of a medium-fine leafing aluminium pigment (STANDART® Aluminium Powder PC 20)

Extrusion of metallic pigments

In this process, the pigment together with all the other components of the powder coating is fed via a premixer into an extruder and the blend is extruded. Extruders are machines in which the mixture is melted and homogenised. Pigments are dispersed completely, if this has not already occurred in the premixer. The gap between the extruder screws is many times larger than the diameter of the metallic pigment particles so that deformation of the pigment relatively rarely occurs.

The extruded product is then grinded. In the mill, shear forces applied to the metallic effect pigments are much higher than in the extruder. In the grinding process, the average particle size of coarse metallic pigment is reduced to 30 to 40 µm. If, for example, pigments with an average particle size of above 40 µm are used, the process leads inevitably to a reduction in particle size and a narrowing of the particle size distribution of the pigment (Figure IV-1.1.3.4). This is directly visible in a reduction of brilliance and destruction of the sparkle effect (cf.

p. 25) of the coarse pigments. Since fine pigments also undergo this reduction in size and deterioration in appearance, brilliant mass tone silver coatings cannot be manufactured by extrusion. Consequently this method is used to manufacture mainly powder coatings containing coloured pigments with limited metallic effect. Bright and brilliant metallic effects are also difficult to achieve in this type of production process because the possibilities for plane parallel alignment of the effect pigment parallel to the substrate are very limited. However, an advantage of this of method is full overspray recycling.

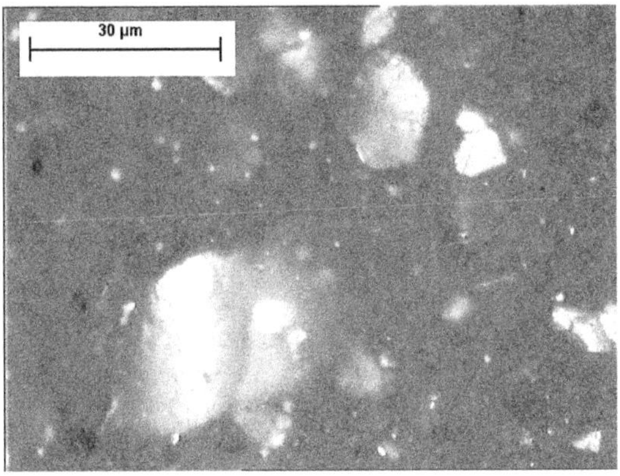

Figure IV-1.1.3.4: Aluminium pigments damaged by grinding

Pigment preparations in a pelletised form can be used instead of pigment powders. They have a pigment content of approx. 95% with the remainder consisting of a binder/additive mixture. The low binder content of the pellets does not affect the overall property profile of the powder coating system. The advantages of pellets over powders are, that they do not dust when mixing into the powder coating, they provide easy handling and dosage, and they do not contaminate the working environment with metal powder dust. In the past, only non-leafing metallic effect pigments were commercially available as pellets. Recent developments have, however, resulted in products (e.g. PowderSafe® from Eckart) with pronounced leafing characteristics which now exhibit significantly higher brilliance than previous pigment preparations.

Bonding

In the thermo mechanical Bonding Process, the pigment is "fixed" to the powder coating particle, enabling the pigment to be significantly better dispersed than in the Dry-Blend Process (cf. Figure IV-1.1.3.5).

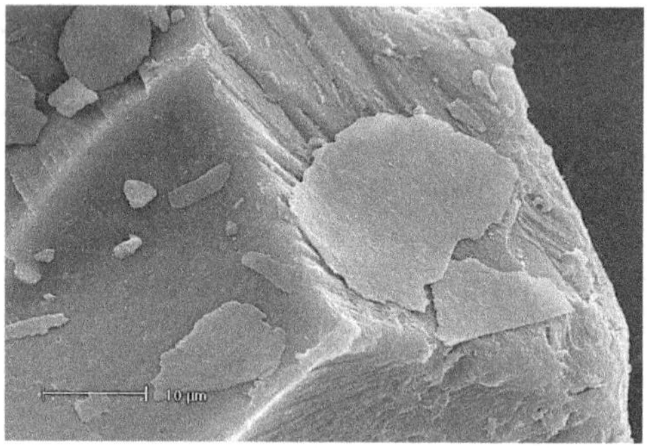

Figure IV-1.1.3.5: Aluminium flakes bonded to a powder coating particle

The process can only be carried out in manufacturing equipment with special safety facilities such as inert atmosphere and effective explosion proofing. The better dispersion of the metallic effect pigments in the binder matrix is particularly obvious in case of leafing pigments. Compared to the Dry Blend, some 30% less pigment is required to achieve a similar bright metallic effect and the appearance of the coating is significantly improved. Figures IV-1.1.3.3 and IV-1.1.3.6 show for purposes of comparison similar powder coating formulations, one produced by Dry-Blending, the other by the Bonding Process.

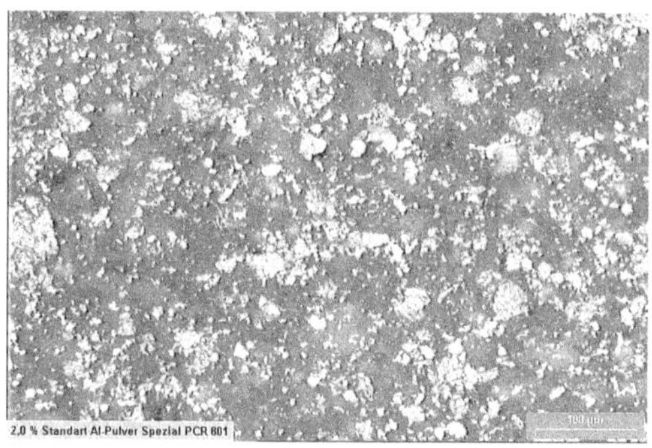

Figure IV-1.1.3.6:
Powder coating manufactured by the Bonding Process with a pigment content of 2.0% of a medium-fine leafing aluminium pigment (STANDART® Aluminium Powder PC 20)

Permanent adhesion of the aluminium pigment to the powder coating particles results in completely recyclable metallic powder coatings. Despite this, it is general practice not to allow the level of recycled powder to exceed 30% in total. Bonded powder coatings are expensive to produce because of the complex process.

Only a limited amount of pigment can be bonded to the powder coating particles because of the limited free surfaces of the binder and the aluminium flakes. As an example, a transparent powder coating with a density of 1.2 g/cm³, a mean particle size of d_{50} 35 µm, and (to simplify this model) a spherical particle shape, has a resulting free surface of 0.14 m²/g. With a coated medium-fine, non-leafing aluminium powder (e.g. STANDART® PCR 501 aluminium powder from Eckart) with a density of 2.7 g/cm³, a mean particle size d_{50} of 20 µm and a pigment thickness of 0.3 µm on a surface of 1.24 m²/g results in a theoretical pigment loading of 11.5% in the binder matrix. This theoretical value is approximately 2.5 times higher than the practical maximum level. Typical pigment loading in practice is 4.5%. Moreover, from experience of operating the Bonding Process, it is known that not every individual pigment particle is actually bonded.

Pigmentation level

Apart from the guidelines discussed for the Bonding Process, the pigment loading in powder coating is generally dependent on the desired effect and the properties of the basic powder coating. In order to achieve full opacity with a pigment particle size below 20µm, the pigment concentration can be lower than as if using a coarser pigment. A pigmentation level of 3.0% of an aluminium pigment with a mean particle size d_{50} of 17 µm results typically in a completely opaque coating, whereas the same pigmentation level using a pigment with a d_{50} value of 47 µm results in a non-opaque sparkle effect finish. The general rule is that the pigment content can be reduced, when using a finer metallic pigment. For leafing effects, considerably less effect pigment is required than for non-leafing effects. Leafing pigmentation varies between 0.5% and 1.5%, calculated on the total formulation.

Flop

The colour of the powder coatings base also affects the flop characteristics of the metallic powder coating (Flop, cf. p. 26). When viewed face-on, the effect is not affected very much, however colour differences become fully visible if the coating is inspected at small angles. Colourless, transparent powder coatings systems provide a darker flop (cf. Chapter II.2: Manufacture and properties of aluminium, gold-bronze and zinc pigments"), while pigmented bases generally exhibit a less pronounced (light/dark) flop.

Over pigmentation

Over pigmentation results in matt, damageable and uneven surface characteristics. Moreover, the robustness of the application process may be negatively affected. When using a powder coatings base containing colour pigments or extenders, lower levels of metallic effect pigment should be chosen than as if working with clear bases. In any case, over pigmentation should be avoided.

Application methods and their effects on coatings

Application processes

Powder coatings are applied using either Corona or triboelectric charge. The advantages of the Corona charging are higher application rates, high versatility and many years of experience with this system. Triboelectric spraying has, however, become increasingly popular with growing demands on flow characteristics and is also advantageous in case of a Faraday Cage. Whereas in the Corona process, negatively charged particles are produced, the triboelectric process results in positive charges on the particles. In the Corona application process, the appearance of the powder coatings film varies within narrow limits as a function of the voltage and actual film thickness. The characteristics of the triboelectric process are different. On the one hand, coatings applied by this technique appear much darker and, on the other, the electrical charge can break down very rapidly during the application process, resulting in the powder coating particles no longer adhering to the substrate.

General comments

During the application of powder coatings, following phenomena occur which directly influence the appearance of the finish: The powder coating material is conveyed from the storage container to the object using compressed air which mechanically accelerates the powder coating/pigment particles. A powder cloud is then formed on the deflector plate. Because of their platelet shape, non-bonded metallic effect pigments exhibit different flight characteristics than the rather spherical binder particles. Metallic pigments "move about" rather uncontrolled in the spray booth atmosphere and then deposit frequently as the final layer on top of the powder coating (cf. Figure IV-1.1.3.2).

Subsequent effects on the coating

It is therefore very possible that the Dry Blend mix of a non-leafing pigment can result in a more brilliant and more metallic appearance than the comparable Bonding formulation. In this case, "more metallic" means a stronger flop when viewed face-on compared to the flop angle. In general, the Bonding Process also leads to pigment orientation on top of the coating surfaces, (statistically there is a certain possibility that a bonded metallic pigment will end up lying on the surface). However, the amount of metallic effect pigment near to the surface in bonded powder coatings is much less than in powder coatings made by the Dry Blend process. The pigment particles oriented on the surface may lead to microroughness, which is clearly visible in a reduction of gloss. Even in case of a high gloss base as an example, the manufacturing of a full shade silver may still result in matting of the surface.

The pigment particles lying close to the film surface are also more likely to suffer from degradation. Moisture, chemicals and mechanical stresses affect the metallic pigments more than as if they are lying embedded deeper in the binder matrix. It is therefore not meaningful to assess overall stability of a metallic pigment without taking into account the way it is bound in the binder. Figure IV-1.1.3.7 shows various applied powder coatings systems with 3.0% of a coarse, coated, non-leafing aluminium powder (STANDART® Aluminium Powder PCR 214 from Eckart). The level of degradation of the metallic pigment depends on the orientation of the aluminium pigment in the powder coating.

Figure IV-1.1.3.7: Appearance after exposure to water condensation: The level of degradation of the metallic pigment depends on the orientation of the aluminium pigment in the powder coating.

In general, durability tests according to the Constant Climate Condensation Test DIN 50017-KK or ISO 6270 (humidity cabinet test: 40 °C, 100% atmospheric humidity) affects aluminium pigments significantly more than exposure to salt spray in the Salt Spray Test according ISO 9227. The reason can be explained as follows: degradation of the pigments in the presence of water is the result of oxidation. Aluminium reacts with oxygen to form aluminium oxide.

$$2\,Al + 6\,H_2O \longrightarrow 2\,Al(OH)_3 + 3\,H_2\uparrow \xrightarrow[-3\,H_2O]{} Al_2O_3$$

During weathering, "degradation" occurs, catalysed by traces of impurities, such as iron or silicon in the aluminium flake. The higher the original purity of the aluminium, the better is the resistance and durability of the metal effect pigment. The durability of aluminium pigments can be further improved, by surface-coating with organic or inorganic compounds. Examples for this are the SiO_2-coated STANDART® PCR and polyacrylate-coated STANDART® PCA pigment powder modifications, as previously described.

Aluminium pigments, coated with acrylic resin are fully compatible with conventional powder coating systems unlike "GMA-Systems" (see above). Coating the aluminium pigment also offers specific advantages during application: aluminium is electrically conductive, the coating, in contrast is non-conductive and significantly reduces discharging of the particles. The risk of voltage break-down caused by formation of a conductive aluminium cloud is therefore minimised.

Safety instructions for aluminium pigments in powder coatings

Detailed advice on safe handling of aluminium pigments in powder coatings is presented in Part VI.

Copper and gold bronze pigments in powder coatings

Besides the general properties of metallic effect pigments, i.e., aluminium and gold bronze flakes, additionally specific characteristics for gold bronze and copper pigments are now discussed. Gold bronze pigments, in as far as they are used in powder coatings, are copper or copper alloy pigments (cf. Part II, Chapter 2.) As described previously, gold bronze pigments are manufactured as leafing grades in the Hametag process. Their thermal stability is limited. The bronze flakes are fairly reactive, i.e. copper oxide is easily formed by reaction with oxygen. The resulting metal oxide is darker in colour and can result in a colour shift, in unfavourable conditions. At higher temperatures, as for example in stoving ovens, the oxidation process on such particle surfaces is significantly accelerated.

Recently developed SiO_2 encapsulated pigments for use in powder coatings (e.g. Goldbronze RESIST and DOROLAN® from Eckart), offer improved protection against oxidation. Such pigments are already in commercial use. Because of the encapsulation, they exhibit non-leafing properties. The applied coatings are fairly resistant against any staining. Stability is further improved by using an encapsulated gold bronze DOROLAN® instead of Goldbronze RESIST Pigment. The performance of both products is compared in Table IV-1.1.3.1 which shows the colorimetric values after 5 minute exposure in a nitrogen oxide atmosphere.

Pigment type		L/25° angle	L/25° angle	L/25° angle	E
Reference:	Resist, non treated	92.6	41.7	70.3	–
Probe:	Resist, 5 min. NO_x-exposed	77.6	32.7	70.1	17.49
Reference:	Dorolan®, non treated	95.8	46.1	69.2	–
Probe:	Dorolan®, 5 min. NO_x-exposed	95.3	42.4	69.9	3.77

Table IV-1.1.3.1: Comparison of the stability of various encapsulated bronze gold pigments

Suitable gold bronzes were used with a mean particle size (d_{50} value) of 17 µm and pale gold colour:

However, the low resistance of Goldbronze RESIST and DOROLAN® to sodium chloride (Salt spray test to ISO 7253) limits their use. This class of pigments can therefore generally only be recommended for indoor use and even then perspiration from hands can cause changes in colour due to degradation. Overcoating with a suitable clear coat can enable these systems to be used in outdoor applications. With specific treatment of the SiO_2-coated pigment surface by special additives (so-called gold bronze PC RESIST-types), the leafing properties originally lost by the encapsulation process, can be recovered. The pigments are now again oriented close to the surface, with effects more brilliant and pronounced, but also more sensitive to touching and consequently degradation. Regarding these properties, gold bronze PC RESIST-types are classified somewhere between straight leafing gold bronzes and real non leafing/coated products.

Table IV-1.1.3.2 shows a comparison of the resistance to condensation of various grades of gold-bronze in the shade rich pale gold. All pigments in this comparison have a mean particle size (d_{50}-value) of 8 µm. The table shows test results for state-of-the art SiO_2-coated gold-bronze pigments (RESIST), and DOROLAN® gold-bronze manufactured by the new improved coating process. Further it contains data for the quasi-leafing post-treated RESIST pigment (PC RESIST) and post-treated DOROLAN®, all supplied by Eckart.

Hours	RESIST (non-leafing)	DOROLAN® (non-leafing)	PC RESIST (quasi-leafing)	Treated DOROLAN® (quasi-leafing)
0	0	0	0	0
168	1.4	0.5	6	1.6
336	1.8	1.0	15	2.6
504	2.0	1.6	17	2.7
672	2.3	2.0	19	3.2
1008	2.4	2.4	21	3.9

Table IV-1.1.3.2: Colour differences ΔE (measured with: Minolta CM-508i/ D65/10) in the condensation test

The data shows clearly that only moderate resistance to acids and alkalis can be expected when gold-bronzes are used outdoors. However, exposure to (UV-) light imposes no restrictions.

Formulation advice

Compared to aluminium pigments ($\rho(Al) = 2.7$), gold-bronze pigments exhibit a considerably higher specific gravity ($\rho(Cu) = 8.9$, $\rho(Zn) = 7.1$). Therefore in a binder matrix with equal pigmentation of metal pigments by weight, there are considerably less gold bronze pigment particles compared to the same formulation with aluminium pigments. This must be compensated by using a higher pigment loading in a formulation. A further consequence of the higher density is the weight of large gold bronze particles, which excludes the use of copper- and gold-bronze pigments with a mean particle size greater than 40 µm. The sinking tendency in the booth and the possibility of the pigment "falling off" the panel to be coated, are fairly high. Therefore the formulation of copper and gold-bronze colours with a sparkle effect is very limited in practice.

Overview of metallic effect pigments used in powder coatings

Table IV-1.1.3.3: provides an overview of the metallic effect pigments which are used in powder coatings.

Pigment type	Orientation characteristics	Average particle size (d_{50})	Opacity	Effect
Aluminium pigment e.g. PCS 1000	non-leafing	10 to 20 µm	very good	fine
Aluminium pigment e.g. STANDART® PCR 214	non-leafing	30 to 60 µm	good	sparkling
Aluminium pigment e.g. STANDART® Chromal XV	leafing	5 to 6 µm	excellent	metallic
Aluminium pigment e.g. STANDART® PC 100	leafing	10 to 20 µm	very good	metallic-fine
Gold bronze pigment e.g. Dorolan® 8/0	non-leafing	8 to 14 µm	very good	fine
Gold bronze pigment e.g. Dorolan® 17/0	non-leafing	17 to 40 µm	good	sparkling
Gold bronze pigment e.g. PC Resist Rotoflex®	leafing	8 to 14 µm	excellent	fine
Gold bronze pigment e.g. Dorolan® 17/1	leafing	17 to 28 µm	very good	sparkling

Table IV-1.1.3.3: Metallic effect pigments in powder coatings

Limitations of metallic powder coatings compared to liquid paints

Coloristics

Powder coaters are increasingly faced with the request of reproducing liquid paint effects in powder coatings. Contemporary metallic powder coatings are reaching their coloristic limits in the brilliant colours required in the automotive, audio, computer and mobile phone industries. The reasons are firstly, the differing pigmentation levels in the dry film and secondly, the significantly inferior orientation of the metallic pigment particles in powder coatings. The photomicrographs in Figures IV-1.1.3.8 and IV-1.1.3.9 provide a comparison of the surfaces of full shade silver automotive finishes produced by liquid and powder coating.

Figure IV-1.1.3.8: Liquid paint full shade silver finish

Figure IV-1.1.3.9: Powder coating full shade silver finish

Abrasion resistance

A frequently encountered problem is the limited abrasion resistance of powder coatings containing metallic effect pigments. This is sometimes termed "blackening" and is caused by abrasive substances attacking pigments on or near to the surface of the coating. The effect is greater the higher the pigment content and the poorer the wetting of the pigment in the powder coating system. Even bonding of the pigment does not improve the situation. The problem can be slightly attenuated by adding Teflon-containing waxes. Abrasion resistance can be subsequently improved by applying a clear coat, although this is usually not interesting commercially. If this option is considered, however, care should be taken to ensure that the clear coat is applied only after the pigmented base coat is fully cured.

In a Dry-in-Dry process followed by co-stoving, disorientation of the pigments would occur, leading to greying. In the Dry-in-Dry process, the powder coating is applied to the substrate and a clear coat sprayed on immediately. Both coats are then stoved together.

Outlook – coloured aluminium pigments

Innovations in the field of coloured aluminium pigments have lead to new coloristic possibilities. In this new class of aluminium pigments, colour pigments are fixed on the surface of the flakes (cf. p 65 ff). The coloured aluminium pigments can be incorporated into clear bases or coloured powder coatings. The resulting coatings exhibit a high and metallic chroma with optical depth effects, which can not be achieved in a coloured powder coating without coloured metallic effect pigments. The coloristic effect can be further enhanced by overcoating with a clear coat.

Guiding formulation for RAL 9006 (white aluminium) indoor		Rec. No. 41-90001	ECKART
1.	Epoxy resin *	48.0	
2.	Polyester resin **	48.0	
3.	Byk 368 P	1.0	www.byk-chemie.com
4.	Benzoin	0.5	

mixing of pos. 1–4, extruding and subsequent milling

5. PCS 1000	2.5	www.eckart.net

Dry-blending or bonding of pos. 5 into the basic powder coating

* Epoxy resin:	Type 3 – 3,5	
	Araldite GT 6063 or	www.huntsman.com
	D.E.R. 663 UE	www.dow.com
** Polyester resin:	Acid value (mg KOH/g) 70 +/- 5	
	Crylcoat 340 or 314 or	www.surfacespecialties.com
	Uralac P 4127	www.dsm.com

Guiding formulation for black antique finish	Rec. No. 41-90002	ECKART
Epoxy resin *	47.5	
Polyester resin **	47.5	
Mowital B 60 H	1.8	www.kuraray.de
Neocryl B 725	0.2	www.noveon.com
Carbon Black FW 200	1.0	www.degussa.com
PowderSafe ...	2.0	www.eckart.net
	100.0	

Resins must not contain flow aids.

* Epoxy resin:	Type 3 – 3,5	
	Araldite GT 6063 or	www.huntsman.com
	D.E.R. 663 UE	www.dow.com
** Polyester resin:	Acid value (mg KOH/g) 70 +/- 5	
	Crylcoat 340 or 314 or	www.surfacespecialties.com
	Uralac P 2980	www.dsm.com

1.1.4 UV coatings

Fundamentals

Coatings systems, which can be cured by high-energy radiation such as UV, are a comparatively recent development. Their success is based on the almost complete absence of solvent emissions on curing. Radiation-curing systems are therefore considered ecologically sound. Up to now, UV coatings have played a relatively small role in the industry. In 2002, their share of the overall coatings market in Europe was 5% [1]. Regarding the binder technology, acrylic systems dominate the UV market with 75% market share, they are followed by polyesters with about 23% [2].

High energy radiation spreads linearly in space. Objects with a large flat surface are therefore ideal for UV curing. All angled surfaces, cavities and complex-shaped objects are more difficult to reach. In these cases the coating application requires numerous UV irradiation sources for complete curing. The requirement of a flat surface is met ideally in print applications. UV print processes have therefore been commercially established longer than UV coating processes. Typical applications of UV coatings are furniture finishes [1] and coating of recording media, (CDs). Flat surfaces (usually wood) are coated with clear coats which are easily cured with UV radiation. This technology is also excellently suited for coating flat plastic substrates [1]. A further important aspect is the very short curing time of just a few seconds, which immediately permits further processing or packing. This is a major advantage when coating mobile phone- or computer cases.

Curing of metallic paints

Inside the paint film, metallic effect pigments act as small mirrors for the high energy UV radiation which therefore cannot penetrate through to the substrate. This can result in inadequate curing of the metallic paint unless it is suitably formulated (see below). Printing inks are applied in very low film thicknesses and are therefore cured without any problems. In the thicker coats of UV metallic paints, complete curing can only be achieved by the use of special high efficiency initiators and catalysts. In practice, a combination of initiators is used in UV metallic paints in order to guarantee a coating completely surface-dry and through-dry. Adequate curing can also be achieved by applying several thin coats, which are cured after each application. Although this process is more costly, it results in a coating with a better property profile. Recently, good curing results have been achieved with the dual cure process in which the UV binder is, as usual, first cured by UV radiation. As the binder also contains isocyanate and hydroxyl groups, they can then be post-cured by heat treatment. Sometimes, the heat treatment is carried out prior to irradiating with UV.

Solvent-free metallic paints

Solvent-free metallic paints are widely available. They contain almost no volatile solvents but only the binder and reactive solvent. Both components react with each other during UV irradiation to form a film. The metallic effect pigments are more or less uniformly distributed in the paint. After application, however, there is only a small likelihood that they will orientate themselves parallel to the substrate. Compared to low-solid coatings with high levels of volatile solvents and a good opportunity for the flakes to orientate, UV system effects appear rather grey and coarse. The use of orientation additives, such as cellulose acetate butyrate (CAB), is excluded, because this would dramatically increase the amount of organic solvents. For effect optimisation, coarser metallic effect pigments (sparkling grades) or grades with a narrow particle size distribution such as silver dollars should therefore be used.

Solvent-based metallic paints

Particularly in Asia, the mediocre orientation of metallic effect pigments in solvent-free coatings has lead to the use of UV coatings with solvents for coating plastic substrates such as mobile phone, television and computer cases. After application, the organic solvent evaporates and the coating is then cured with UV irradiation. Although this method is at odds with the initial concept of an environmentally-friendly VOC-free application process, there are major advantages in the excellent property profile, outstanding appearance and comparatively rapid processability in the production process. All aluminium pigments, which are also used in low-solid coatings, are suitable. There are no limitations in average particle size or particle size distribution as in the solvent-free UV metallics. In the case of gold-bronzes, coated grades (see below), which offer adequate stability, are used.

Metallic effect pigments used

If the rules mentioned above are considered, both aluminium and gold-bronze pigments can be used in UV coatings. Both have a certain potential for reaction which often causes uncoated standard aluminium and gold-bronze pigments to catalyse the polymerisation of the binder components. In such cases, UV metallic paints can react completely after just a few hours of storage. The use of coated metallic effect pigments such as silicate-coated gold-bronze STANDART® Resist or Dorolan, grades or aluminium pigments such as STANDART, PCR and STAPA® Hydrolan in powder form (STANDART® aluminium powder PCS), permits the formulation of UV metallics which are stable against polymerisation. Polymer-coated aluminium pigments can also be used. All the metallic effect pigments mentioned (manufactured by Eckart) are supplied as powders and therefore do not introduce any solvent into the UV coating.

Recommendations for formulating

Particular care must be taken when dispersing the pigment: coated metallic effect pigments have a high specific surface area and are often not uniformly distributed in the poor-wetting UV binder. If necessary, a suitable wetting agent should be used. With gold-bronzes, precautions should be taken to prevent settling as these pigments have a high specific gravity.

Water-based UV coatings

A recent development in the field of UV curing finishes are water-based systems in which the binders cure by UV light, reacting just like in traditional UV coatings although they are dispersions in water. Drying takes place in two steps: in the first by evaporation of the water and, in the second, by curing of the binder using UV irradiation.

Water-based UV metallics

In water-based UV systems, it is necessary to use stabilised aluminium pigments to prevent the pigment, firstly, from reacting with water and generating hydrogen, and secondly, from reacting with the reactive UV binder. Aluminium pigments stabilised with a SiO_2-encapsulation, such as STAPA® Hydrolan from *Eckart*, have proved particularly suitable (cf. p. 90 ff). The encapsulation around the aluminium pigment resists attack by water and from the UV binder, and permits the formulation of paints with a long shelf life.

An advantage of water-based UV metallics over solvent-free systems is the significantly better pigment orientation and the resulting excellent optical effect. It can be further improved by addition of water-based wax emulsions.

Gold coloured water-based UV coatings

The stabilised gold-bronze pigments STANDART® Resist and Dorolan® are not stable enough for use in water-based UV coatings. Consequently, the aluminium pigments mentioned must be combined with transparent organic pigments to imitate gold and copper colours.

References

[1] *Beck, E.*: UV-Härtung unter Kohlendioxid – für jedermann einsetzbar. *Farbe & Lack*, March 2002, pp. 98 to 107

[2] *Balmer, B.*: UV market: the latest trends. *European Coatings Journal*, 2001, Vol. 10, pp. 406 to 408

Guiding formulation
Solvent based UV-metallic-basecoat
for plastics

Rec. No. 85-09001

ECKART

Stapa Aluminium Paste	5.00	www.eckart.net
Methoxy propyl acetate	30.00	
predisperse		
Butyl acetate 98/100	45.00	
Lucirin TPO	0.40	www.basf-ag.de
solve		
Laromer LR 8863	18.50	www.basf-ag.de
Irgacure 184	0.35	www.cibasc.com
Irgacure 500	0.35	www.cibasc.com
Byk 333	0.40	www.byk-chemie.com
	100.00	

stir for 10 min. at appr. 1000 rpm.

Viscosity:	app. 11 sec. DIN cup 4
Flash off:	10 min.
Curing:	UV curing time is depending on equipment
Application pressure:	4 bar

Guiding formulation Water-based metallic UV basecoat		Rec. No. 86-09002	ECKART
Component A:			
Stapa IL Hydrolan ...		5.00	www.eckart.net
Disperbyk 191		0.50	www.byk-chemie.com
Component B:			
Wasser demin.		2.50	
Butyl glycol	mix and stir for 5 min	2.50	
n-Methylpyrrolidon	at appr. 1500 rpm	5.00	
Lucirin TPO-L		0.50	www.basf-ag.de
Component C:			
Alberdingk Lux 399		60.00	www.alberdingk.com
Alberdingk Lux 101		22.00	www.alberdingk.com
Irgacure 500	mix	0.75	www.cibasc.com
Irgacure 184		0.75	www.cibasc.com
Dow Corning 57, 10% in H_2O		0.20	www.dowcorning.com
Byk 333		0.30	www.byk-chemie.com
		100.00	

First paste Component A with B. After that put Comp. C to the mixture of A and B and stir for 10 min at appr. 800 rpm.

Viscosity: appr. 19 sec, DIN cup 4 mm
Drying: 10 min. at 80 °C
Curing: the curing time is depending on equipment
pH-value: 8.1 to 8.3
Application pressure: 4 bar

1.2 Applications

1.2.1 Automotive coatings

Modern automotive coatings consist of four layers, each one of which fulfils a specific function. Together they are about as thick as a human hair. The first layer, the primer, protects the metal from corrosion. The second layer, the filler, smoothes out the smallest unevenness and so forms an optimal base for the thin colour layer which, with pigments and effect agents, gives the optical effect to the coating. The final layer is a clear coat which protects the whole system and provides optimal gloss.

Why silver?

Silver, produced using flake shaped aluminium pigments, is the colour which best accentuates the shape of the car body. No other colour reflects the light better than silver and silver is one of the colours least affected by dirt. Its metallic shimmer enhances the shape and plays iridescently with light and shade. Silver has the image of modernity and innovation. Although silver vehicles comprise well over 30% of sales in Germany and Japan, the multiplicity of the modern silver palette leaves plenty of room for individuality as there are around 6000 known formulations of grey and silver tones. In addition to the technically cool colour mixtures which underline their modern hi-tech effect by a light blue nuance, there are real alternatives in the so-called soft silver tones. A combination of aluminium pigments and micro titanium produces a "cool" colour with a golden-yellow colour reflection.

Regional differences

In spite of the globalisation and internationalisation of the market, trends and quality demands differ greatly depending on region (Figure IV-1.2.1.1).

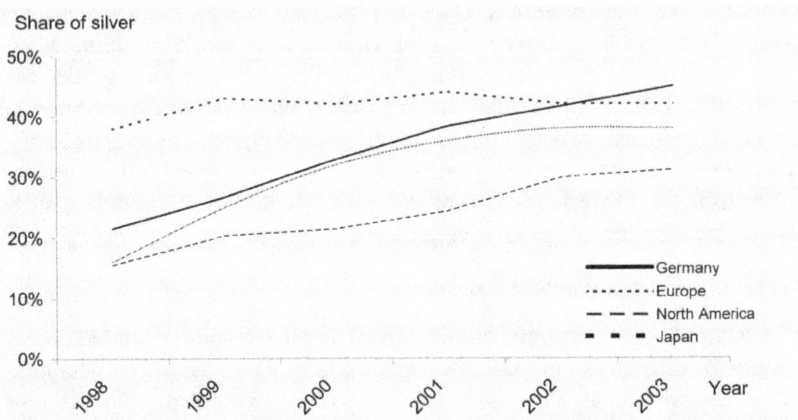

Figure IV-1.2.1.1: Colour trends in automotive application: silver

Europe is a design-driven market and demands on the optical properties of automotive coatings are very high. The customer expects high-gloss, uniform coatings in brilliant colours. North America is very different; the market and further paint developments are clearly dominated by technology. In general, coarse and sparkling metallic effects are in fashion. This naturally fits the usual high-solids coatings with their significantly poorer orientation of the metallic pigments. Brilliance, brightness, colour purity and degree of gloss play a reduced role. The popularity of colours and effects depends not only on the customer's personal taste but the colour must also suit the type of vehicle and its design.

Coating production

For all the effects described, the same products are manufactured to the same standards worldwide. In recent years a concentration process has taken place among the automobile manufacturers and as a result a similar process has occurred among automotive coatings producers. Today there are only a few manufacturers but they operate globally. For example an automotive finish produced in Europe can be applied in Asia or coatings made to the same formulations will be used for vehicles on sale in America and Asia. For metal flake pigment manufacturers, this means that the same pigment must be made globally available in the quantities required. For cost reasons, many coatings manufacturers only store small quantities of raw materials and demand just-in-time delivery. The required products must also be delivered at the required time – not too early and certainly not too late – and this makes major demands on logistics.

Pigment grades in Europe

Modern silver dollar types are used in Europe. Because of their high metallic brilliance and good flop, bright and brilliant colour shades can be produced. Fine, opaque products are preferred ($D_{50} < 15$ μm). To obtain the highest possible gloss, coarser fractions are removed (limiting grain size 25 μm). Aluminium wheel rims and plastic wheel trims are often coated with coarse sparkle grades and this allows the desired glitter effect to be easily achieved. Sometimes finer, opaque grades are added in order to obtain adequate coverage with thin paint film thickness. Increasing costs in recent years have caused automobile manufacturers to move to more cost effective cornflake grades with narrow particle distribution. These products have a good price/performance ratio.

Pigment grades in North America

For technological reasons, North American automobile manufacturers make somewhat different demands on metallic effect pigments with moderate appearance. Such products were used in Europe 10 or 15 years ago but are now only found in coatings systems used for refinishing.

Pigment grades in Asia

In countries with intense sunshine (e.g. South China, India, Korea), coarse sparkle types are extremely popular (often mixed with fine opaque metallic effect pigments to achieve good covering) because they have a beautiful glitter effect. In Japanese-dominated markets, up to 3-5 years ago white was predominant. In recent years (c.f. Figure IV-1.2.1.1) silver has experienced a real boom with fine and highly brilliant grades being preferred. Compared with western types they have a weaker flop. The reason for this is that Japanese automobile manufacturers use metallic effect pigments which were developed for the electronics industry (mobile phones, computer parts). They are highly opaque, brilliant and fine but only have a low flop.

The usual pigment content is between 3 and 4% for full shade silver coating and 0.2 to 0.5% for dark metallics. For all colours the coatings are formulated so that two spray passes give complete coverage.

Technological development

Quality consistency

The term automotive coatings immediately brings to mind production line painting, mainly of car bodies, specifically with steel as substrate. Nowadays, not only auxiliary parts such as bumpers are made of plastic but also wings. Optically they are completely integrated into the car body but are coated offline with different coating systems using different application techniques to those used for the metal body. Modern vehicles have a nil-gap concept, that is metal and plastic components touch each other completely. Demands on colour matching have therefore grown considerably and visible colour deviations are not acceptable. This puts enormous demands on the coatings industry and the pigment manufacturers. Different coating systems on different substrates and under different conditions of application must result in exactly the same shade.

For metallic effect pigment manufacturers this means making available a broad and diverse range of products to meet the demands for optimal orientation and dispersability in different coating systems. Additionally, high quality must be consistently maintained within narrow tolerances for every product. Practically, these high demands for quality consistency involve incorporating the intended metallic effect pigment into the base coat and applying it in a test situation. The applications are measured with a multi-angle spectrophotometer against an agreed standard. Acceptable batches have a colour deviation at the critical 25° angle of $\Delta E^* < 1$.

Purely visual assessment, as often used for metallic effect pigments, is no longer appropriate. In addition to the appearance, the particle size distribution is also specified within very tight tolerances. Nevertheless, the solids (non-volatile) component of the pigment paste is of decisive importance. Too high or too low a proportion of pigment in the paste leads to incorrect metallic pigment concentration in the coating system and thus to different shades. This can be a problem especially in coloured metallic finishes.

Colour variations should be kept to a minimum by the control measures described. This is particularly critical for certain colours such as yellow. Water-based coatings make particular demands as even small deviations in formulation can drastically alter the shade. This means that testing of metallic effect pigments in water-based binders is often client-specific. Additionally, testing of gassing stability must also meet the requirements of the individual client.

Technology of the coating systems

Colours and coating technology, coatings chemistry and application techniques have all changed markedly in recent years. The change-over of automotive coatings to environmentally friendly water-based systems has certainly made the greatest demands on aluminium pigment manufacturers. As it reacts with water, aluminium cannot be used in water-based systems without protection, and stabilisation of the pigment is absolutely essential (cf. Part IV, Chapter 1.1.2).

The complete range of stabilised metallic effect pigments is used. For mild automotive base coats, metallic effect pigments stabilised by additives are used as this offers economical formulation. Intercoat adhesion between the water-based base coat and the clear coat plays an important part in the property profile. Therefore metallic effect pigments coated with different levels of additive are used. Basically, the smaller the amount of additive, the smaller the side effects, such as problems with intercoat adhesion – however the higher the risk of gassing. Aggressive binders require correspondingly well-stabilised pigments. From a historical point of view, the first pigment generation comprised aluminium pigments which had been chromated. Until recently, they were unsurpassed for universal compatibility and high gassing stability. Thanks to their inert surfaces they have practically no side effects in the binder matrix.

However the heavy metal chromium found on the pigment is environmentally dubious. It is true that as a trivalent compound it is not directly biologically dangerous, but in spite of this it is present in the coating. Regulations governing recycling of old cars require that as high a proportion as possible be recycled, and chromium is therefore no longer acceptable.

Here the SiO_2-coated types (cf. p. 88 ff) offer a true, heavy-metal free alternative They are just as gassing resistant and universally applicable as their predecessors. While water-based systems are established and state-of-the-art in Europe, North

America has followed the path of high-solid systems. The amount of organic solvent in the formulation has also been drastically reduced there.

Additionally solvents which at present are believed to have only a limited ecological effect are used. These include, for example, "HAPS-free" solvent (HAPS = hazardous air pollutant solvent), which are classified as not harmful to the atmosphere under defined test conditions.

The demands on aluminium pigments in high-solids are quite different from those in water-based coatings. In principle the usual aluminium pigments are used without additional post-treatment in high-solids. However, it must be noted that the visual impression is different from that in low- or medium-solid coatings or water-based systems. The high solid content of the coating causes the orientation of the aluminium pigment to change: they appear to be coarser, darker on face and whiter on angle, and the overall appearance is somewhat restless.

Coarse pigments, with average particle sizes D_{50} larger than 25 µm, are preferred. Even with only modest orientation they still produce attractive effects. The same is true for the use of aluminium pigments with the narrowest possible particle size distribution. On the one hand they lead to excellent appearance and on the other hand absorb little binder because of the relatively small pigment surface area. Consequently the viscosity of the high-solid coatings also remains within an acceptable range.

In this context non-degrading Tufflakes® already described (c.f. p. 127 ff) are ideal for high-solid systems. They are relatively coarse and thick and have a very narrow particle size distribution and therefore have only minimal effect on the viscosity. However their low hiding power often requires them to be combined with opacifying silver dollar or cornflake aluminium pigments.

New demands are now being made on the aluminium pigment manufacturer: While coatings in solid colours have for a long time been applied electrostatically, the last layer of the metallic basecoat is always applied pneumatically to obtain the required bright, brilliant silver colour. To reduce costs and perform the coating process more quickly, efforts are being made to achieve the same colours as previously but using electrostatic application. The relevant details are given on pages 143 ff.

Car refinishing

Car refinishing paints are available commercially as Ready-Mix systems (that is ready-to-use coatings) and mix systems. For the latter, demands on the constancy of all technical properties of the metallic effect pigments are extremely high. With only 6 to 8 types of aluminium pigment up to 40,000 metallic shades must be manufactured by carefully worked-out formulations. Apart from criteria already stated, it is important that the variation in the shade of automotive refinishes from batch to batch should be as small as possible. The refinish should match exactly the original colour of the undamaged parts of the vehicle. Because of the complicated drying and application conditions there are only a certain number of water-based refinishing systems available commercially. Their application amounts are nothing like that of the OEM automotive finishes. Both Ready-Mix systems and mix systems must be kept for years in closed containers and there must therefore be absolutely no gassing. This is only possible if coated aluminium pigments are used. Additive-stabilised grades cannot offer this reliability.

Plastic accessories

When considering automotive interiors, there are scarcely any limits on the imagination. However the demands on stability and durability of the coating are completely different from those of car body finishes so that there is a multiplicity of choices in the area of styling. Not only the optical properties but also the haptics play a decisive role, yet the interior of a vehicle should produce an atmosphere of well-being. It is also important that the textiles used should match in texture and colour. There is a certain trend towards decorative coatings (cf. p. 131 ff).

Basically, the styling of automotive interiors is more conservative and discrete than that of electronic articles. The automobile interior should be elegant and innovative. Flashy styling of large areas would be off-putting. Moreover, automobiles (and therefore their interiors) are used for significantly longer than mobile phones or computers and a certain timeless design is necessary. External plastic parts of the vehicle are usually coated in the same colour as the chassis.

Gold bronze pigments

Gold bronze pigments have found no application in automotive industry. There are several reasons: the high density means that settling in the circulation line is not controllable. The same is true of the high reactivity, i.e., the inadequate chemical stability of this group of pigments. Combinations of aluminium pigments with organic colour pigments are available as simpler, cost-effective and more practical alternatives, e.g. yellow-orange or gold colours.

Guiding formulation
High solid basecoat (OEM)
baking system

Rec. No. 50-09022

ECKART

Stapa Metallux 2156	7.5	www.eckart.net
Butyl acetate	8.5	
paste		
CAB 551-0.2, 20% in butyl acetate	30.0	
Joncryl 507	30.6	www.johnsonpolymer.com
Cymel 303	10.5	www.cytec.com
Dowanol PMA	6.0	www.dow.com
Butyl acetate	3.5	
MAK	2.4	www.eastman.com
Dynapol Catalyst 1203	1.0	www.degussa.com
	100.0	

Thinner:	xylene/butyl acetate 1:1
Spraying viscosity:	20 sec/DIN cup 4 mm
Dry film thickness:	app. 15 µm
Flash off time:	30 min.
Stoving conditions:	30 min. at 120 °C

Guiding formulation **Metallic basecoat** **baking system**		Rec. No. 76-09014	ECKART
a)	Stapa IL Hydrolan 2192 No. 55900/G	4.65	www.eckart.net
	Water demin.	4.65	
	Disperbyk 192	2.50	www.byk-chemie.com
	Byk 345	2.50	www.byk-chemie.com
b)	Bayhydrol LS 2952	27.05	www.bayer.com
	Water demin.	13.51	
	Butyl glycol	1.35	
c)	Dimethyl ethanol amine 10% in H_2O demin.	4.50	
	Viscalex HV 30	2.52	www.cibasc.com
	Water demin.	13.70	
	Cymel 325	3.37	www.cytec.com
d)	Butyl glycol	1.35	
	Water demin.	4.45	
e)	Water demin.	13.90	
		100.00	

Mix all raw materials one after the other as per the following sequence:
b – c – d – a and e

Viscosity: with H_2O demin. to 30 sec./DIN cup 4 mm
pH-value: app. 8
Drying: 30 min. at 120 °C

1.2.2 Decorative coatings

In our modern industrial design for consumer goods, the importance of surface quality is continuously growing. The application of a high-quality surface coating makes substrates such as wood or plastic appear to be sophisticated and unique: accentuating the shape and adding value to it. The demand for high-quality finish is often based on the fact that more and more products almost copy each other in shape. In order to give their products unique character and to differentiate them in the market, designers select an unmistakable surface finish. A few years ago as an example, televisions were only available in anthracite black and a number of solid colours. To create an eye-catching effect, metal pigments were co-extruded with the plastic. Nowadays, we are finding chrome-effects or glittering silver finishes – providing the impression of quality and luxury.

The haptics – touch and feel

A very important design aspect is to impart a high-quality character to the specific object by "optical material imitation". The development of modern surface finishes often goes a step further than that and also includes a variation of the haptics, i.e. coatings with a certain feel to the surface. A somewhat cold and metallic surface can now be coated with soft-touch coatings – making the object seem soft, velvety and pleasant to the skin. Within these systems, it is important that haptics and appearance correlate. It is not desirable to apply a frosty, cool, metallic look as a soft-touch formulation. Coatings with a harmonious blend of haptics and appearance are used in equipment which is often touched or held by hand - such as mobile phone or computer mouse.

Accentuation of the substrate properties

A further design strategy is to emphasise and accentuate the characteristics of a substrate. For example, the structure and grain of wood allow unique effects to be produced with suitable coatings. This phenomenon is based on the specific orientation of the metal effect pigments in the grain, so that it is strongly emphasised. In this area, the furniture industry has discovered the almost endless possibilities of aluminium and gold-bronze pigments. Whether emphasising natural wood grain or textured MDF sheets, special effect finishes are central to create 'ambiance'.

Protection of substrates

When using effect coatings, practical aspects may not be forgotten. Metallic effect pigments can significantly contribute to prolonging the life of the substrate. In the case of wood they prevent from crack formation. Metallic effect pigments also protect against UV radiation and slow down the degradation of binders. Plastics are preserved from bleaching and embrittlement, and metal is protected against corrosion. In the following, tried-and-tested formulations of stylings with metallic effect pigments are presented. They can be formulated in an almost endless variety with liquid coatings; this field will now be described in detail. The colour photographs in the middle of the book and the relevant specialist chapters show further areas of application.

Effects with aluminium pigments

Combinations with organic colour pigments

Transparent organic pigments are commonly used to achieve coloured metallic effect coatings. This type of mixture maintains the brilliance of the metallic effect pigment together with a high colour depth. Either coarse or fine aluminium pigments can be used. With coarse aluminium pigments, a highly-brilliant, sparkle effect is obtained with a marked flop – also surfaces appear textured. On the other hand, fine-particle effect pigments produce a very homogeneous and soft surface appearance. The flop transforms smoothly into the darker coloured metallic shade. In general, these effects are the more brilliant, the more transparent (fine) the organic pigment is. The opposite is the case for aluminium pigments. The coarser the metal particles and the narrower the particle size distribution, the higher the brilliance.

Combinations with inorganic pigments

A further possibility of formulating new types of attractive effects is by colouring the coating system with inorganic pigments. Contrary to the light-absorbing nature of organic pigments, inorganic pigments produce their effect by scattering light. By addition of a small amount of titanium dioxide for example, the flop from bright to dark can be gradually reduced without significantly affecting the metallic character. In this way the brightness for special formulations can be increased and more homogeneous surfaces obtained. However, the addition of large quantities of inorganic pigments leads to a strong reduction of the metallic effect. Moreover, the coatings have a milky, matt appearance.

Effects with oxidised aluminium pigments

With oxidised aluminium pigments (cf. p. 60 ff) it is obvious from their unique champagne colour, that they are different from usual aluminium pigments. Most effects obtained by combination of organic pigments and oxidised aluminium flakes are not likely to be matched by using conventional aluminium flakes. The coarsest grade, Aloxal® PM 4010 (Eckart), is characterised by high brilliance and strong flop when formulated in dark shades. The fineness of Aloxal® PM 2010 and Aloxal® PM 3010 grades (Eckart) results in surfaces with a very homogeneous effect. Oxidised aluminium pigments are ideal for warm shades such as gold or copper colours. Yellow, orange and red organic pigments partly lose chromaticity in combination with normal aluminium pigments. With Aloxal® the colours remain clear, pure and brilliant. Other mixtures with coloured pigments are also interesting as Aloxal® generally shows a stronger flop than the corresponding base pigment. The strong bright-dark effect makes this pigment especially suitable for round objects. Combinations with carbon blacks produce completely new shades which are favourably distinguished within the usual anthracite colour area.

For coatings with pronounced flop, coarse pigments are preferred as they exhibit a stronger flop. For combinations with conventional aluminium pigments, fine-particled products are recommended. In such mixtures a silver colour with a touch of gold is obtained, it has a very strong flop and seems particularly impressive. In order to make sure not to lose the effect of Aloxal®, high-quality silver dollar pigments should be used in combination. Depending on the mixing ratio, all shades and colour nuances from silver to champagne can be produced.
Other colouristic possibilities of oxidised aluminium pigments include the inexpensive production of anodised finishes with a soft texture. Furthermore, by addition of aluminium oxides or waxes to the formulation, an object can be made to appear as if it was sandblasted. Not only the appearance but even the feel can be imitated with remarkable accuracy: the surface feels as rubbed by fine or coarse sand paper, depending on the type of additive used. In addition to the colouristic possibilities, oxidised aluminium flakes also offer advantages in application. They have unusually high circulation line stability which can otherwise only be achieved by using non-degrading flakes. This high stability is useful in applications with strong shear forces. Further to that, the chemical stability is excellent, which is an advantage in the one-coat systems that are commonly used in the decorative field.

Effects with white coloured aluminium pigments

Completely new effects can be obtained with this new generation of coated aluminium pigments. White-coloured aluminium pigments are characterised by an extremely bright appearance when viewed at face angle. If the coating is observed from another angle, a darker flop is seen which goes into a slightly bluish colour. Alucolor® white (Stapa Alucolor® 3004 weiß – Eckart) offers a combination of brightness and metallic appearance not previously available. The big advantage of this pigment class is shown in combination with various organic colour pigments. While strong, intensive shades are sometimes difficult to obtain with mixtures of conventional aluminium pigments and organic pigments, high Chroma colour effects can be obtained with white aluminium pigments. Because of their simple formulation, these pigment mixtures are particularly valued with problematic colours such as red or yellow.

A further possibility of application lies in formulating frost-effect coatings. The characteristic bluish-white colour of this pigment, which already appears to be frosty in masstone, can be combined with micronised titanium dioxide to obtain even more spectacular ice-like coatings. The very good orientation of the white coloured aluminium pigments produces a very homogeneous surface. This advantage is particularly evident with difficult tasks such as imitating steel or tin.

Effects with green, blue or gold coloured aluminium pigments

In order to obtain an extreme depth of colour with metallic coatings, coloured aluminium pigments are preferred. The inherent colour of the effect pigments gives the impression of being able to see "into" the coating. The most extreme colour flops are obtained, if coatings with coloured pigments are applied on substrates in respectively complementary colours. As the effect pigments are relatively coarse, only the sparkle effect of the pigment is visible face-on. If the coating is inspected from a different, smaller angle, the effect of the pigment disappears and the colour of the substrate dominates. With this type of semi-transparent coating it is essential that application is accurate and cloud-free. As extreme colour changes are not always desired, single coating is also used. By combination with suitable organic pigments of the complementary colour, a clear but somewhat weaker colour flop can be obtained. Furthermore, these pigments, at a concentration of 1 to 3% on total formulation, are excellent for providing a subtle extravagance to the coating. For example, in deep black coatings, the coloured aluminium pigment is almost invisible; however in direct sunlight the coating suddenly glitters with lively colour and produces the desired "wow" effect.

Effects with gold-bronze pigments

The colours of gold-bronze pigments vary from copper to greenish gold. Pure, brilliant gold, yellow, orange, red and copper metallic colours can easily be achieved. These fire-coloured shades produced by oxidation, such as fire-coloured red, deep gold and lemon yellow – offer new design possibilities. As it is the case with coloured aluminium pigments, gold bronze pigments open up a whole new range of colour combinations. Using organic coloured pigments, highly brilliant flop effects can easily be produced. The intensive gold colour is favourably affected by micronised titanium dioxide as the flop effect is extended. Special encapsulation of gold bronze pigments makes it possible to use them in a great variety of coating systems. This encapsulation also improves the chemical resistance. However, their tendency towards oxidation limits the use of gold bronze pigments in industrial coatings and they are therefore restricted to interior applications and consumer electronics.

Effects with PVD pigments

PVD pigments (cf. Part III, Chapter 1) are extremely fine aluminium effect pigments, which are manufactured by the very costly procedure of vacuum metal deposition. Unique effects can be realised with PVD pigments such as Metalure® (Eckart). The most common use is in chrome effect coatings. When formulating a base paint containing PVD pigment, the following principles must be kept in mind: The ratio of binder (solid) to Metalure® (solid) should not exceed 3:1 otherwise the desired effect cannot be obtained. At spray application, excess pigmentation gives the impression of a fine-particled but standard aluminium pigment. The coatings formulation is always a compromise between perfect appearance and the chemical and physical resistance of the coating. Keep the amount of binder in the formulation as small as possible. The more pronounced the film shrinkage immediately after application, the better the orientation of the PVD pigment. In standard formulations the dry film thickness is just three to five micrometres. As solvents, mixtures of high and medium volatility should be used. Because of the very low film thickness, only very smooth, textureless, glossy substrates should be used. Basecoats with these special pigments are unable to cover unevenness or defects in the substrate. Very beautiful effects are obtained if sprayed opaque or semi-opaque on top of highly-gloss black primer.

PVD pigments can easily be combined with transparent organic coloured pigments. There are practically no limits to styling creativity. However, to avoid orientation problems and associated problems with the overall effect, the coloured pigments must be very fine dispersed. Otherwise the PVD platelets “lean on” coarse organic colour pigments which consequently reduces the effect. Seen face-on, this type of combination with transparent pigments is silver metallic and the flop, which represents the colour pigment, is very dark at a flat angle of observation. By far the most brilliant shades with a previously unachieved depth are created in combination with dyestuff solutions. The only restricting factors are the limited light fastness of the dye and the potential migration of the dye into the coating. Because of the low film thickness and the resulting low mechanical and/or chemical resistance, it is common, to apply an additional clear coat on top of the effect coat.

A further area of application for coatings with PVD pigments is the counter coating of glass or transparent plastics such as acrylic (PMMA). The PVD pigmented coating is applied to the reverse of a transparent substrate to produce an almost perfect mirror. For this type of application the amount of binder must be smaller than in the coatings discussed above. The adhesion on plastics can be improved by the right choice of solvent. As soon as the effect appears matt or streaked, dilution is too great. This special effect pigment, correctly formulated, is a flexible, economic and environmentally-friendly alternative to matt metallization.

Special effects

Combination with micronised titanium dioxide

If aluminium pigments are combined with micronised titanium dioxide, a champagne or light gold shade is obtained when viewed face-on. The sophisticated effect flops into a milky blue if inspected almost parallel to the surface. This "frost effect" is already widely used in the automotive sector and gaining popularity in the field of consumer electronics. Oxidised aluminium pigments such as Aloxal,, with already champagne-like colour, are supporting this effect to such an extend, that the colour produced face-on is almost gold. At other angles of observation, a remarkable blue-shift is produced which leads to excellent flop effects in combination with organic coloured pigments. The colour differences achieved with gold bronze pigments are strong at different angles of observation. The micronised titanium dioxide increases the gold colour and the flop. These interesting results are only obtained if the following three conditions are fulfilled:

- good orientation of the metallic effect pigment
- narrow particle size distribution of the metallic effect pigment
- distinctive flop of the metallic effect pigment

The strongest effect is obtained with coarse silver dollar pigments. The ratio of titanium dioxide to aluminium should lie between 1:3 and 1:0.5 depending on the desired effect. If over-pigmented with titanium dioxide, a more matt, less-metallic effect is produced. In general such combinations result in a silver or gold-shimmering appearance face-on and a matt flop into cool blue. In addition to straight mixtures with micro titanium dioxide, combinations with additional organic colour pigments can also be recommended. This significantly strengthens flop behaviour makes the actual shade even more attractive.

Combinations with pearlescent pigments

Silver coloured pearl effect pigments have a greater brilliance and brightness than aluminium pigments. To exploit the great brilliance of the pearls and the high covering power of the aluminium pigments, the two are often used in combination. The most attractive combinations regarding brightness and brilliance are obtained with fine-particled silver dollar pigments and fine-particled silver pearls. The low opacity of the pearlescent pigments plays a major role in the overall impression. The higher the opacity, the more metallic, brilliant and bright is the effect. Combinations of interference pearl pigments and metallic effect pigments are increasingly used as they allow soft and sophisticated flop effects with different colour nuances at different angles. To cover the gold-, beige- and sand-coloured effect coatings area, earth-coloured pearl pigments are combined with gold-bronze and oxidised or conventional aluminium pigments. These mixtures also offer a great variety of surface design possibilities.

Combination with semi-transparent organic pigments

There are numerous new types of organic pigments which offer a greater hiding power than conventional organic pigments. In combination with metallic effect pigments these show a significant behaviour: metallic effects are stronger viewed face-on, imparting a greater metallic character to the coating, but if the viewing angle towards the surface is decreased, the metallic effect almost completely disappears. Metallic pigments with pronounced flop are of advantage here. Combinations with additional organic pigments and/or micronised titanium dioxide are also possible. They further improve the optical effects and increase the options for creativity.

Combinations with dyestuffs

In combination with metallic effect pigments, dyes produce a very high brilliance. Additionally to that, a depth of colour is obtained, which is not achievable with organic pigments. However, compared with organic coloured pigments, low light fastness of the dyestuff is a problem with this type of effect coating. Combinations of dyestuffs with PVD pigments should be particularly mentioned here, as these effects are almost on the same level as coloured metalized objects. Depending on the application, in some cases it can be useful to colour the clear coat rather than the base coat.

Sandpaper effects

Sandpaper effects are obtained with decorative coatings which have a certain roughness from addition of structuring agents. Appearance and feel should be correlated accordingly. When using coarse metallic effect pigments, coarse texturing agents should also be used and vice-versa. In this way the complete range of sand blasting effects including coarse sandpaper effects can be produced. Special aluminium pigments are utilised to make sure the visual appearance correlates with the feel. Alongside this classic combination, the visual effect can be strengthened by mixing with pearlescent pigments and organic coloured pigments. For everyday objects which are frequently used, haptics play an increasingly important role.

Soft-touch effects

"Soft-touch" coatings feel very soft and velvety to the touch. This effect is produced by very coarse metallic effect pigments. The coating system is very important as it determines the basic properties. The soft-touch effect is often determined by the thickness of the film which is usually matt and has no flop. Even silk-like effects are possible with special pigments. Micronised silicone resins can be used in combination with metallic pigments. These resins produce an effect similar to soft-touch with the advantage of higher metallic brilliance at lower film thickness.

Marble effects

Marble effects are particularly attractive. These effects can be produced with special stencils. Several contrast pigmented layers of paint are applied on top of each other within a special stencil technique. This generates complex optical structures with a certain multidimensional aspect. Novel effects can be produced by using metallic effect pigments for one or several of the top coats. The first layer (e.g., red) is partly overlaid by fine, bright, metallic, shining veins which stand out under suitable illumination. The various top coats can be applied in any order; however the best effect is obtained if a metallic coating layer is applied first of all. The process is suitable for surfaces where artistic accents are required, for example, decorative frames.

3D-effects

The so-called 3D-effects are visually particularly attractive. Contrary to textured or hammer effect paints (see below) the surface is completely even and glossy. Such an effect coating can be overcoated with a clear coat to increase the 3D-effect. The landscape of virtual planes and spaces suggests the impression of spatial depth. Here also, the combination of illumination and observation angle is crucial for the ever-changing variations of the visual impression. As with the high-grade steel effect, high-quality, fine metallic pigments are required. The 3D-effect is generated by controlled disorientation of effect pigments directly after the application. The pigment loading in the metallic effect coating is lower than usual (approx. 1%), as a certain amount of transparency emphasises the texture. The actual colour of the underlying primer (e.g. grey, beige) is also contributing to the result, as background for the overall impression.

Hammer finish paints

Hammer finish paints are coatings systems which can be applied as a single coat to metal objects or machinery. In general they are inexpensive and help to cover up unevenness or imperfections in the substrate. These systems have a randomly textured surface structure which is obtained by incompatibilities of high molecular weight silicone oils. The aluminium pigments emphasise the visual expression of this incompatibility. Alongside silver coloured hammer finish paints, coloured hammer finish paints based on gold-bronze pigments are also commercially available. The latter allow production of very brilliant gold shades. In both cases, mixtures with organic pigments are also possible as this contributes to the attractiveness of the overall impression.

Flop effects

For soft flop effects aluminium- gold-bronze-, oxidised aluminium- or coloured aluminium pigments are usually combined with pearl interference pigments. A further possibility is the combination with semi-transparent, organic coloured pigments or micronised titanium dioxide. The most spectacular flop effects are obtained with expensive special effect pigments (Colour-Shifters) or require multicoat application.

Textured coatings

Textured coatings are coating systems which have a more or less uneven surface. This effect is obtained by incompatibility and high thixotropy. In general, silicone oils are not used for this. Very coarse metallic aluminium or nacreous pigments are suitable and even combinations with organic coloured pigments are possible. With the growing role of haptics, the interest for textured decorative coatings also increases.

Crackle finishes

Crackle finishes are made particularly attractive by using metallic pigments. To obtain the desired crackle effects, two contrasting top coats are applied, e.g. black on gold. Crack formation is based on rapid film drying with consequently high forces within the film structure. Strong film stress arises from different elasticity of the coating systems. Within the first minutes, the top coat cracks and forms a regular crack structure in combination with the partially exposed primer. Crackle finishes are mostly nitrocellulose lacquers which dry very quickly by solvent evaporation. They are excellent for a large number of substrates.

Water-drop coatings

Water-drop effects require a multi-layer structure. One coating film always contains an effect pigment. Aluminium- and gold-bronze pigments can be used. In one layer only water is applied to a dried film. It is important that only a few water droplets touch the object. The water must then evaporate. Immediately after this a further coating layer must be applied. Generally the whole is finally covered with a protective varnish. Most interesting are coloured water-drop effects. Such coatings are already finding application in the furniture industry.

1.2.3 Industrial coatings and functional applications

General considerations

There is nowadays an increasing trend of using metallic-effect pigmented coating systems not only in the automotive and consumer electronics industries but also in various areas of industrial coatings. In addition to aesthetic considerations, it is often functional aspects which make the use of metallic-effect pigmented coatings attractive.

Following areas of application deserve specific attention:

- can-coating
- coil-coating
- marine paints
- anticorrosive paints
- architectural paints and roof coatings
- general industrial coatings
- engineering plastics

Functional properties of metal effect pigments are based on their morphology, as well as on their physical and chemical characteristics. Examples of functional applications of aluminium pigments are in marine- and anticorrosive paints. Due to their platelet-structure their barrier-effect provides corrosion inhibiting effects. In architectural paints and roof coatings the most important feature is the fact that metallic-effect pigments reflect both heat and light. Other functional characteristics, such as electrical conductivity or thermal durability are useful for formulation of conductive coatings or temperature resistant systems.

Can-coating

In can-coating, use of non-toxic aluminium pigments, permitted for direct contact with foodstuffs by FDA §21 Nr. 175.300, is a key feature. Any irregularities which may appear on the metal substrate are effectively concealed by the metallic effect pigments. Usually fine aluminium pigments with average particle size less than 15 µm are used. This is important as the coating thickness is only between 8 and 12 µm. Many products are specifically labelled as food-grade, which is, permitted for contact with food. In some cases, coatings are required, which are chemically very stable. Only surface-coated aluminium pigments can satisfy this strict demand. The most suitable types are special SiO_2-coated aluminium flakes (cf. p. 88 ff). Originally developed for water-based coating systems, they can also be used successfully in solvent-based systems. Gold-bronze pigments are generally not permitted for coatings with direct because of the heavy-metal content.

Coil-coating

Metallic-effect pigments are highly valued in industrial coating of metal coils (roller coating). In one or two coat application, metallic-effect-pigmented top coats (18 μm) can be applied on top of an epoxy primer (5 μm) at a conveyer speed of about 120 m/min without any problem. Typical binder systems for durable façade-elements (exterior use) are PVC, PVDF and PUR. With these, non-leafing metallic-effect pigments are well embedded in the binder matrix. Further to their aesthetic function the opaque metal pigments reflect radiation in the range of visible- and UV-light and consequently protect the binder polymer matrix against photolytic degradation.

Regarding the optical properties, these metal pigments don't have pronounced flop effects. Coil-coating applications for packaging generally involve a low film thickness. Therefore, it makes sense to prefer pigments where coarse particle fractions above 45 μm or even 25 μm have been removed. These adequate pigments result in smooth and even coil-coating films. For formulating coil-coating applications (interior or exterior panels) almost the whole range of aluminium pigments is available.

For exterior use when high water and moisture resistance is to be achieved, polymer coated flakes are among those which can be employed (cf. p. 88). SiO_2-coated products offer particularly high chemical resistance. High colour reproducibility is demanded by manufacturers of coatings used for façade components, for both interior and exterior use. Very tight colouristic specifications for the aluminium pigments are the basis for producing façade panels that are also very consistent in colour. Even panels from different production batches can now be mounted next to each other. As gold-bronze pigments are not sufficiently stable, they are only very rarely used in coil-coating application.

Marine paints

Aluminium pigments are used in marine paints because of their barrier effect (see below). This is strongest for coarse grades. Fine-particled pigments have short diffusion paths for water vapour and are therefore less suitable. The same considerations also apply to coatings for "off-shore" structures such as oil drilling platforms and containers.

Gold-bronze pigments were long used as sources of copper in anti-fouling coatings on ships' hulls. Slow corrosion liberated copper ions which strongly inhibited growth of algae. However, modern anti-fouling coatings largely avoid the use of metallic-effect pigments.

Anti-corrosive coatings

Aluminium- and gold-bronze pigments are not true anticorrosive pigments. Nevertheless, aluminium flakes as leafing grades are popular as they are UV-reflectors and protect the coating layer from decomposition by chalking (see below.). Non-leafing pigments are mainly used as barrier pigments against humidity and therefore they slow down the corrosive degradation of steel substrates (see below). A combination with micaceous iron oxide is usual. Here the aluminium pigment brightens up the total colour impression. Further to that, the aluminium platelets integrate ideally with the micaceous iron oxide to strengthen the barrier protection. True anti-corrosive paints are containing active anti-corrosion pigments or zinc dust respectively zinc flakes. They are fully discussed in Chapter 5 of Part IV. Combinations of zinc and non-leafing aluminium pigments lead to coatings which can be successfully welded. This has advantages in building large structures such as bridges.

Architectural coatings and roof coatings

Architectural coatings and, importantly, roof coatings, mainly exploit the ability of aluminium pigments to reflect radiation (see below). In this way the reflective coating protects the building from excessive temperatures.

General industrial coatings

Metallic-effect pigments are gaining importance in industrial coatings generally. Traditionally, metallic effects were mainly used for metallic effect-pigmented hammer finishes for machinery and machine parts. Today, modern design principles are more and more applied in all industry areas and therefore strongly involving effect coatings. Medium-fine cornflake aluminium pigments with wide particle size distribution find many applications in grey-metallic industrial colours. These are specified by the RAL-Organisation as RAL 9006 ("White aluminium", exclusively based on aluminium pigments) and RAL 9007 ("Grey aluminium", that is aluminium pigments combined with micaceous iron oxide). Both aluminium types, leafing and non-leafing, are recommended. Further to these silver RAL-colours, the coatings industry develops individual colour variations on demand. Recently, metallic effects are being employed in coating wood and furniture, in bicycle paints and increasingly in coating plastics used with electronic equipment.

The excellent hiding power of aluminium pigments – combined with metallic brilliance and flop – are the main reasons why metallic effect finishes are used increasingly for coating consumer and capital goods. The brilliant metallic effect imparts a sophisticated appearance to products and increases their perceived value (cf. Part IV, Chapter 1.2.2).

Plastic coatings for mobile phones, digital cameras, television sets etc. are predominantly applied as single layer coatings. In addition to being decorative, the coating must provide high chemical resistance. Effect pigment suppliers offer various coated aluminium pigments to respect these high demands. Furthermore, the metallic-effect pigments must be fully embedded in the binder matrix. As a test method for this, a piece of adhesive tape is applied to the coating and then sharply pulled away, no metallic-effect pigment must be visible on the adhesive tape ("Tesa"-Test). Of course there must be adhesion to the plastic substrate. This is additionally tested using a cross-hatch adhesion test to DIN 53151 or EN ISO 2409.

Methods of application

The formation of the metallic effect depends very much on the method of application. Various application techniques are suitable for metallic paints.

Pneumatic spray coating application

Metal effect pigmented paints can be applied by spray application without any problems. The use of different spray methods may lead to differences in results. Spray coating with an air pressure of 4 to 5 bar produces excellent results. The pneumatic spray coating application is the preferred method for all industrial and automotive coatings with metallic effect pigments although the application efficiency rate is rather low.

Airless application

Metallic effect pigmented coatings can be applied without additional compressed air (airless). The atomisation of the coating into fine droplets is guaranteed by a high system pressure. As the layer thicknesses of the applied coatings are higher than when using pneumatic atomisation, the orientation of the pigments and thus the optical development of the metallic effect is not comparable to pneumatic applications. This means that coarser metallic effect pigments are usually used compared with those used in pneumatic spray coating – to provide attractive appearance. The choice of metallic effect pigments is often limited by the cost. The fields of application (see below) only allow the use of economic products with large particle size distribution (cornflake grades) and moderate optical effect. Leafing grades are frequently used.

In spite of the limited metallic effect, the airless technique is often used in industrial coating, especially for large areas such as containers or ships. The reasons for this are maximum application efficiency, higher film thickness in one pass, and the possibility to process also paint systems with high viscosity. The saving in solvent compared with pneumatic spray coating is also of importance.

Electrostatic spray application (ESTA)

On ecological grounds, electrostatic spray application is becoming increasingly important. The advantage of this method is that the amount of overspray, that is the amount of spray which misses the substrate, is significantly reduced. Usually the efficiency rate of application is over 70%. In practice, this technique is combined with pneumatic support. There are two methods of electrostatic application: In the first, the paint droplets are produced by a (spray) bell at 45,000 rpm; about 2/3 of the metallic base coat is applied in this way. The remaining 1/3 of the total paint amount is applied in a second step by pneumatic application. Currently this two-step procedure produces the best compromise between degree of effectiveness and surface finish quality. The target is to achieve 100% coating by bell atomisation.

The second method involves formation of paint droplets by a rotating disc. The rotation speed is about 20,000 to 35,000 rpm. Both surface finish and metallic effect are good and it is used amongst other applications for bicycle frames. Both methods involve high voltage of about 75 kV.

In principle all metallic effect pigments can be used for electrostatic spray application, although optically high-quality aluminium pigments with narrow particle size distribution are preferred as they allow good optical properties to be obtained. This application technique is particularly of advantage, whenever high productivity rates are required, e.g. in OEM coatings for automotive. Also in the area of industrial coatings, electrostatic methods are gaining ground.

Roller application

There is no problem applying metallic-effect coatings using rollers and this method is heavily used in coil- and can-coating. However, there are differences in the optical effect depending on the direction of movement, which remain also visible in the finished items.

Dip-coating application

Dip-coating objects in coating baths with metallic paints allows surfaces with a good metallic effect to be produced. In cases where a higher film thickness is required, e.g. with candles, decorated flowers or other decorative items, the dipping procedure can be repeated several times.

Brush application

Metallic effect paints can be equally well applied by bushing. The optics is however not as uniform as with spray application because the brush direction directly affects the pigment orientation and therefore the effect produced. A certain manual skill is required to produce coatings with rather homogeneous metallic effects by brush application. However, visible brush tracks, caused by irregular orientation of metallic pigments, cannot be entirely avoided. Examples of areas where this technique is used are water-based decorative paints.

Functional applications of metal effect pigments in industrial coatings

While the quality of the visual effect produced by metallic-effect pigments mainly depends on colour, dispersability and particle size distribution as quality factors, chemical and physical properties are the most important criteria for use in functional applications. Changing application profiles and growing technical demands on the coatings have led to continuous improvement of metal pigments with advanced characteristics. Especially when talking about durability and chemical resistance of metallic pigments, the demands of the market can be met with sophisticated products. When now these "high-performance" metal pigments are combined with adequate binders for industrial coatings, remarkably high levels of system durability can be achieved.

These important functional properties of metallic effect pigments will be discussed further:

- electrical conductivity
- thermal conductivity
- heat stability
- corrosion protection, electrochemical properties
- reflection of electromagnetic radiation (e.g.: visible light and UV)

Electrical conductivity: conductive coatings

The electrical conductivity of metals is used in electrically conductive varnishes [2], conductive coatings [3, 4] and conductive plastic coatings (e.g. conductive flooring in computer rooms) [5].

The classic pigment for conductive coatings is silver, which is used in high-quality printed circuits or in applications such as heated rear windows of cars. Silvered copper powder is a commercially available substitute for pure silver. Pure copper pigments are also used to a certain extent in conductive coatings [6], but beside sedimentation these have the disadvantages of easy oxidation. The clear advantage from copper over silver is a favourable price. The limitation for copper powders is their corrosion tendency, especially if exposed to higher temperatures.

For a long time nickel powder was favoured for use in conductive coatings as it combines the advantages of the previously considered pigments: good conductivity, chemical resistance, thermal resistance and relatively low price. However, because of the toxicity of nickel powder, its use has become almost completely taboo.

Aluminium can not be used as conductor for these applications, because the individual aluminium pigment particles are surrounded by an oxide layer that is electrically isolating.

Table IV-1.2.3.1 compares conductivities of different metallic effect pigments at different pigmentation loading. As expected, the electrical resistance decreases with increasing pigment/binder ratio (P/B).

Metal effect pigment	p/b ratio 85/15	p/b ratio 70/30	p/b ratio 50/50
Nickel powder ECKA VP/AT	0.02 kΩ	0.05 kΩ	30.0 kΩ
Kontaktargan (silvered copper powder)	0.002 kΩ	0.004 kΩ	0.2 kΩ
Copper powder ECKA CH 516 (without isolating oxide layer)	0.01 kΩ	0.02 kΩ	< 100,000 kΩ
Copper powder ECKA 081 B (without isolating oxide layer)	600 kΩ	180,000 kΩ	< 100,000 kΩ
Copper powder ECKA MP 7200 (without isolating oxide layer)	600 kΩ	21,000 kΩ	21,000 kΩ
STANDART® Copper powder Lack L900 (with isolating oxide- and lubricant layer)	8,000 kΩ	11,000 kΩ	< 100,000 kΩ

Table IV-1.2.3.1: Resistances of metallic-effect pigmented conductive varnishes measured with:
binder: Neocryl B725 (40% in butyl acetate/isopropanol 1:1),
thinner: ethanol/ethyl acetate 1:1, spray viscosity 20 sec DIN 4 mm,
stoving conditions: 30 min/100 °C,
average particle size of all pigments: 20 μm ± 4 μm

Electrically insulating metallic-effect pigments

As already mentioned earlier, aluminium pigments are not suitable as conductors due to their isolating oxide layer. This might imply that they are perfectly suitable for coating television sets – which is not the case. Due to the fact that voltage in the kilo-volt range is applied, aluminium pigmented coatings can't exclude electrical breakdowns completely. Therefore for coating of TV-sets so-called "non-conductive" aluminium pigments are used, which are encapsulated in an electrically isolating acrylic layer (cf. Part IV, Chapter 1.2.2 : Polymer coating).

Thermal conductivity

Metallic effect pigments are also thermally conducting but this effect is scarcely used in surface technology. In the plastics industry it is used as a way of removing heat in, for example, moulds, compound materials and frictional linings.

Heat resistance: heat resistant coatings

Because of their relatively high melting point, metallic-effect pigments also exhibit good temperature stability. Metallic-effect pigments, predominantly coarse-particled leafing grades, are therefore used in heat resistant coatings for heating installations or car silencers (Figure IV-1.2.3.1). Aluminium pigments are mainly used in silicone-based binders where they are resistant almost up to the melting point (660 °C). The use of stainless steel pigments for these applications is negligible. At higher temperatures a non-oxidisable aluminium-iron compound (Al_3Fe) is formed which offers effective protection against further oxidation [7].

Figure IV-1.2.3.1: Exhaust pipe-metallic coated

Corrosion protection: electrochemical properties

Aluminium pigments are contributing to corrosion protection via barrier-effect, e.g. in marine paints [9]. Here the roof-tile-like arrangement of the pigment flakes retards diffusion of water through the coating layer, and therefore also delays swelling and permeation of the coating film (Figure IV-1.2.3.2).

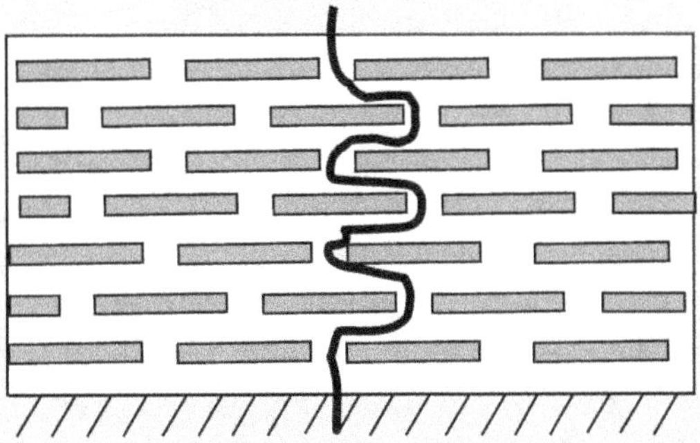

Figure IV-1.2.3.2: Protection by reflection at metallic flakes

Zinc-dust pigments have had an established place in corrosion protection for many decades. This uses the potential difference between iron and zinc to produce an effective cathodic protection against corrosion. The most important areas of application for zinc-dust primers are in coating steel structures (bridges, electricity pylons, etc.), especially those objects which for practical reasons cannot easily be galvanised. In addition to isometric zinc dust, flake shaped zinc products are also available commercially. These offer effective barrier protection as well as cathodic corrosion protection (cf. p. XXf [Part IV, Chapter 5: Modus operandi of zinc flake pigments, Topic: In addition to corrosion protection...]) [8].

Reflection of electromagnetic radiation

The reflecting properties of flake shaped metallic-effect pigments are closely related to the anti-corrosive effect of aluminium coatings. The reflection of UV-light protects the organic binder matrix which lies below the leafing aluminium pigments, from photolytic decomposition (cf. Figure IV-1.2.3.3) [10]. Because of their comparatively low price, coarse-particled products are mainly used. Re-coating must be done regularly to maintain the reflecting properties of the coating. The intercoat adhesion can be ignored for this application. The typical example is roof coating.

The reflection of heat and IR radiation is mainly made use of in tank coatings and, as already mentioned, roof coatings (cf. Figure IV-1.2.3.3) [11]. In addition to corrosion and UV protection, these coatings also inhibit overheating of rooms below them. In this connection aluminium coatings behind radiators are equally effective in reflecting heat directionally.

Reflection of visible light is also important in coatings for screens and weather balloons.

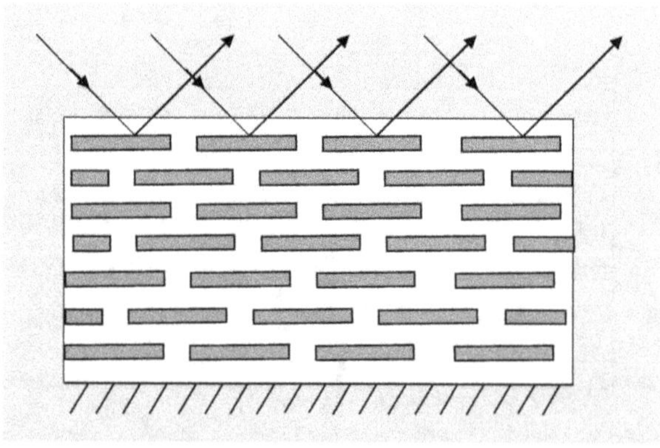

Figure IV-1.2.3.3: Protection by reflection at metallic flakes

Shielding electromagnetic radiation

As already described, metallic-effect pigments function as small mirror-like reflectors for visible light. This raises the question whether electromagnetic radiation in the radio frequency range is also shielded by the platelet shaped particles (EMI-Shielding: Electro-Magnetic-Impact-Shielding). If this would be the case, the transmitting and receiving power of electronic equipment (coated with metallic pigments) could be affected. Thorough investigation with mobile phones have shown that no shielding by metallic-effect pigmented coatings, affecting the transmitting or receiving performance, can be detected [1].

Outlook

Newly developed aluminium pigments, which have exceptional chemical resistance and excellent weathering stability, are opening up new fields of application. Pigments completely encapsulated with a SiO_2-layer (e.g. Hydrolan® from Eckart) are already being successfully used to protect wooden facades. This type of pigment was originally developed for water-based systems but use of isopropanol as paste forming agent also allows it to be employed in solvent-based coatings. The SiO_2-layer provides the exceptional chemical resistance properties. With extremely thin coatings of 25 to 30 μm, wooden facades show no damage at all even after years of outdoor exposure. Wood can thus be protected from long-term weathering using extremely low pigment concentrations (Figure IV-1.2.3.4).

Figure IV-1.2.3.4: Protection by light reflection on a wood substrate (Biberist, Switzerland)

References

[1] Expert report from Prof. Dipl.-Ing. *Peter Pauli* (Abteilung Hochfrequenz und Radartechnik in Neubiberg-München)

[2] *Lendle, E.*: Metallverbindungen in lufttrocknenden und einbrennbaren Lack- und Siebdruck-Bindemitteln. In: Metalloberfläche – Angew. Elektrochemie, 1974, 28, vol. 2, pp. 64 – 69

[3] *Ebneth, H.; Fitzky, H.-G.*: Kunststoffteile mit guter Schirmwirkung für hochfrequente elektromagnetische Wellen. PLASTverarbeiter, 1982, no. 7, pp. 760 – 764

[4] GB 2055386; Acheson Ind., 1980

[5] DE-OS 2428504; Ciba-Geigy, 1974

[6] DE-OS 30228114, Acheson Ind., 1980

[7] *van Oeteren, K. A.*: Aluminiumpulver – Beschichtungsstoffe, Material und Anwendungen. Die Mappe, 1980, no. 6, pp. 374 – 375

[8] *Besold, R.*: Neuartige Korrosionsschutzpigmente auf Basis blättchenförmiger Zinkpigmente. Farbe & Lack, March 1983, pp. 166 – 173

[9] *Funke, W.*: Umweltfreundlicher Korrosionsschutz. Organische Beschichtungen – Probleme und Realisierungsmöglichkeiten. In: Farbe & Lack, February 1983, pp. 86 – 91

[10] *van Oeteren, K. A.*: Aluminiumpulver – Beschichtungsstoffe, Material und Anwendungen. Die Mappe, 1980, no. 6, pp. 374 – 375

[11] *Griffin, C. W.*: Cost savings with heat-reflective roof coatings. Plant Engineering, July 1980, pp. 98 – 102

Guiding formulation **2K-Reflection paint, silver**	Rec.-No: 37-09014	ECKART
Component I:		
Stapa 4 Alum.	12.0	www.eckart.net
Xylene	10.0	
Solvent naphta	2.0	
predisperse		
Dowanol PMA	2.5	www.dow.com
Raybo 41	0.5	www.raimund-mueller.de
Desmophen A 160, 60% in xylene	59.0	www.bayer.com
Component II :		
Desmodur N 75	14.0	www.bayer.com
	100.0	
Thinner:	butyl acetate 98/xylene 1:1	
Spraying viscosity:	18 sec. DIN cup 4 mm	

Guiding formulation
PVC-Aluminium paint, based on Laroflex Rec. No. 30-09011 ECKART

Stapa 5 VS	12.0	www.eckart.net
Solvent naphtha	9.0	
Tetralin	3.7	
predisperse		
Laroflex MP 45, 20% in solvent naphtha	69.7	www.basf-ag.de
Laropal A 81, 50% in xylene	2.8	www.basf-ag.de
Haftharz LTW	1.4	www.degussa.com
Xylene	1.4	
	100.0	

Viscosity: 30 sec. DIN cup 4 mm

Guiding formulation
Roof coating aluminium Rec. No. 30-09021 ECKART

Disflamoll DPK	9.0	www.bayer.com
Barytes heavy spar	36.2	www.baryt.com
Aluminium Paste Stapa 2 Aluminium	4.1	www.eckart.net
Cab-o-Sil TS 720	0.4	www.cabot-corp.com
Methoxy propyl acetate	5.1	
Catalyser DBTL, 10% in methoxy propyl acetate	0.9	www.kever-technologie.de
Bevaloid	0.5	www.cherbsloeh.de
Desmodur LD	2.0	www.bayer.com
15 min under vacuum stirring		
Desmodur E 15	36.9	www.bayer.com
Desmodur Z 4470	1.8	www.bayer.com
Hardener OZ	3.1	www.bayer.com
	100.0	

5 min under vacuum stirring

Guiding formulation
Epoxy-phenolic lacquer with aluminium pigments for can coatings Rec. No. 43-09009

ECKART

Stapa Al-Paste 65%ig	4.50	www.eckart.net
Solvent naphta	8.16	
predisperse		
Epoxy resin* 31,8 % in solvent mixture**	73.00	
Phenodur PR 722, 53 BGB	13.42	www.ucb-group.com
Additol XK 406	0.72	www.ucb-group.com
Lanco flow L or Disparlon L 1980	0.20	www.noveon.com www.kusumoto.co.jp
	100.00	

*** Epoxy resin:**

Beckopox EP 309	www.ucb-group.com
Araldit GT 6610	www.huntsman.com
Epikote 1009	www.brenntag.be
D.E.R. 669	www.dow.com

**** Solvent mixture:**

Methoxy propanol	26.6
Methoxy propyl acetate	13.4
n-Butanol	20.0
Solvent naphta	40.0
	100.0

Epoxy/phenolic ratio: app. 7.65 : 2.35
Solid content (% by weight): app. 33

Film properties on tin plate (cured 10 to 12 min at 200 °C)
Dry film thickness: 5 to 7 µm
Acetone double rubs: 10

2 Print

2.1 Offset

Metallic effect pigments for offset printing

Gold and silver effects in packaging and labels increase the perceived value of a product towards the consumer. Therefore gloss and brilliance are especially of importance. Offset printing is the most commonly used printing process worldwide to achieve eye-catching metallic effects in advertising, packaging materials, brochures, labels and illustrated books.

Offset printing

Offset printing differs substantially from other printing techniques. Image areas and non-image areas on the printing plate are almost on the same level. Depending on the type of plate, the image areas can lie a few thousandth of a millimetre above or below the non-image areas (however this is basically of no importance for the printing process). Colouring and printing in offset technology is possible because the non-image areas are absorbing (water) fountain solutions and thus repel the printing ink. The image areas are absorbing very little fountain solution, they mainly pick up printing ink. The image areas are lipophilic (attract fats) while the non-image areas are hydrophilic (attract water).

Offset ink

The offset printing ink must be formulated to be lipophilic and water repellent. To produce a sharp closed printed image, the ink must also be able to pick up fountain solution transferred to the non-printed areas, i.e. it must be capable of forming a stable emulsion. Figure IV-2.1.1 shows the layout of an offset printing press.

Printing process

In the first step, the printing plate, which is fixed to the plate cylinder, is treated with fountain solution. This further increases the difference in hydrophilic character of the non-image areas and the lipophilic character of the image areas. The printing ink is then transferred via several ink application rollers onto the printing plate. A complex roller system ensures uniform distribution of the ink. The amount of ink transferred can be controlled by the opening of the duct screw adjustments and the oscillation of the inking roller. The maximum wet film thickness of offset ink is around 1 to 2 µm. Offset inks have a very high viscosity, similar to that of honey, which permits high print sharpness.

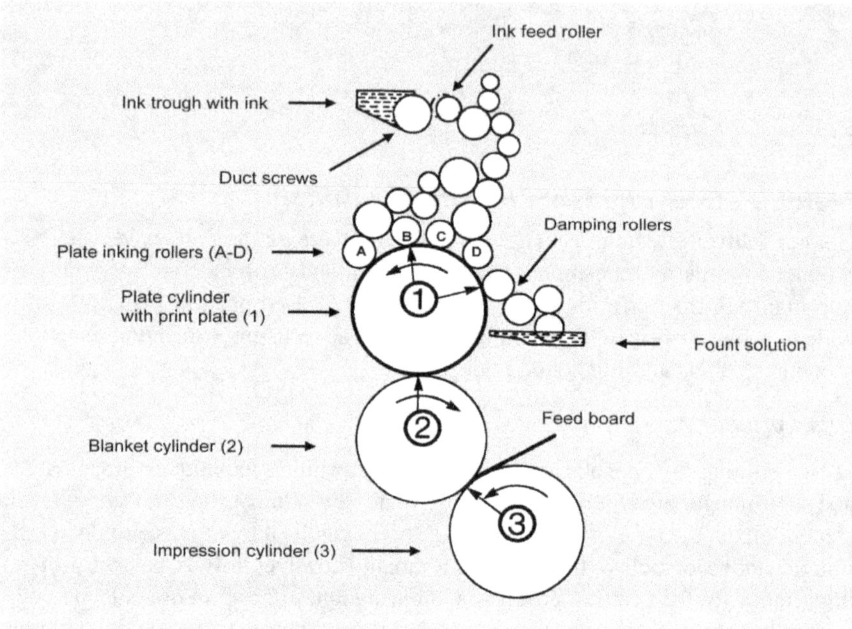

Figure IV-2.1.1: Principle of offset printing

Metallic effect pigments for offset printing inks

Metallic effect pigments for offset inks have to be very fine in particle size to ensure good transfer via the complex roller system inside the printing press and to provide good coverage at low film thicknesses. The range of metallic effect pigments which can be used is therefore limited to the fine-particle end of industrially produced products. Established products, such as Lithoflex® pigments from Eckart, are available with average particle sizes (d50) of 2 to 4.5 µm for gold-bronzes and 4 to 9 µm for aluminium pigments. These grades exhibit cornflake morphology (cf. p. 31) and are manufactured by the Hall wet grinding or Hametag dry grinding processes. Almost all metallic effect pigments used in offset printing inks are leafing. As previously discussed (p. 19), this leads to floating of the pigments on the ink film and thus to greater brilliance of the printed image.

Table IV-2.1.1 shows the metallic effect pigment grades used in offset printing:

Product	Average particle size (d_{50})	Delivery form	Opacity	Brilliance	Effect
Leafing goldbronzes (Standard or leafing-stabilised)	2.0 to 2.8 µm	powder or paste[1]	very good	very good	fine
e.g. Lithoflex® from Eckart	2.5 to 4.0 µm	powder or paste[1]	very good	excellent	fine
Leafing aluminium pigments	4.0 µm	paste[1]	excellent	very good	extremely fine
Leafing aluminium pigments	4.0 to 8.0 µm	paste[1]	very good	excellent	fine
e.g. Stapa® from Eckart	6.0 to 10.0 µm	paste[1]	good	good	fine

[1] high-boiling mineral oils, aromatic-free, used as pasting agent

Table IV-2.1.1: Metallic effect pigment grades

Surface-treated metallic effect pigments

Some of the binders used in offset printing have strong pigment-wetting properties. In order to minimise this effect, surface-modified pigments with improved leafing stability are offered in addition to the stearic acid coated grades (cf. p. 15). This surface modification – a technique used for Lithoflex® XA grades – significantly extends the shelf life of the printing ink and reduces the loss in brilliance caused by pigment wetting. Metallic effect pigments are usually supplied in paste form. In general, the products are supplied in high boiling mineral oils such as PKWF 4/7 af neu, or in fatty acid esters such as Estisol 312 (both from the Haltermann company). The solvent content of a bronze paste is usually about 13%, regular aluminium pastes contain 35% solvent.

Advantages of press ready offset inks for the printer

Nowadays, metallic effects for offset printing are sold in the form of press ready ink. Hardly any manufacturer of printing inks produces metallic effect inks in house. The reasons are obvious: formulation and production of metallic offset printing inks with a long shelf life require own development facilities and special equipment. As metallic effect printing inks represent less than 1% of the turnover of most manufacturers, it is simply uneconomic to manufacture them. Most gold and silver inks are bought in as ready-made products from pigment manufacturers and sold under the ink supplier's own brand name.

The manufacture and processing of press ready offset printing inks are described in the following part.

Press ready one-component offset printing inks

Besides press ready one-component systems, some two-component inks are still available. In these, the pigment paste and the binder are supplied separately and used to mix the ink in a ratio of 1:1 (gold) or 1:3 (silver) just before printing. The disadvantages of two-component systems are the additional work involved and a lower storage stability after mixing. Modern one-component inks are press ready and are in no respect inferior to two-component systems in terms of printability and brilliance. Easy handling and high storage stability are their advantages. Consequently, nowadays one-component inks are used almost exclusively.

Basic principles of formulating offset inks with metallic effect pigments

Metallic effect printing inks consist mainly of two components: the pigment and the binder. While the pigment provides the colour, the binder is responsible for transport through the printing press and for the final adhesion to the substrate.

The essential criterias for manufacturing modern one-component gold and silver printing inks are:

- the right choice of pigment (optimised on the basis of parameters such as leafing/non-leafing, average particle size/particle size distribution and surface modification)
- specially formulated binders which enable an optimum pigment orientation in the ink film
- choice of special polymers considering any potential reactions with metallic effect pigments
- special production equipment which enables smooth pigment dispersion

Taylor-made surface modification of the effect pigments and careful choice of the binder will determine the level of pigment wetting. Proper wetting will ensure good transfer on the printing press combined with good pigment orientation in the ink film. Maximum brightness and brilliance in offset printing are achieved, as described earlier, by the use of leafing pigments. Additionally metallic printing inks are also highly pigmented (see below). Non-leafing systems, on the other hand, exhibit outstanding pigment wetting with a orientation of the pigment within the ink film. Although such systems have reduced metallic effects, they can easily be overprinted and have good finishing properties. However, due to the reduced brilliance they are hardly used in offset printing.

Optimum brilliance

The opacity and brilliance of a printing ink are mainly determined by the rheology of the binder. Suitably formulated ink rheology is essential for optimum pigment mobility after the ink is applied in order to achieve plane-parallel pigment

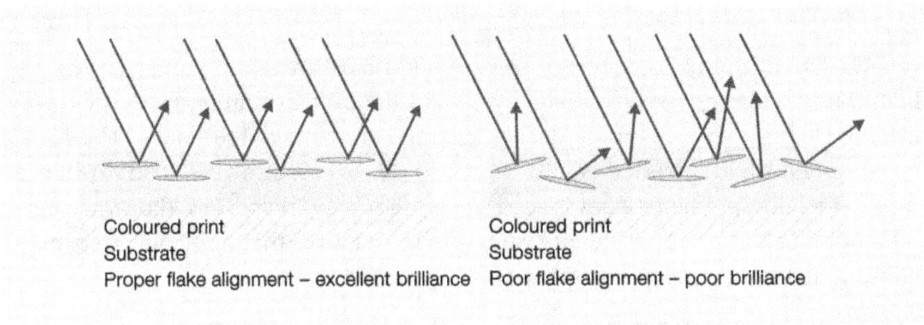

Figure IV-2.1.2: Pigment orientation in a print

orientation. It is essential for the ink to exhibit good flow behaviour on the substrate with the least possible increase in printed dots – this would reduce the sharpness of the printed image (Figure IV-2.1.2).

As a general rule, binders in conventional offset printing inks consist of phenolic-modified rosins or hydrocarbon resins solved in mineral and/or vegetable oils mixed with alkyd resins. A press ready gold offset ink usually has the following composition:

Pigment	45%
Binder	50%
Dryer	2%
Wax paste	3%

Printing inks pigmented with aluminium flakes contain 15% to 18% of metallic pigment and a correspondingly higher amount of binder.

Because of the reactivity of metallic effect pigments, the selection of raw materials for formulating the binder requires special attention. High water content of the raw materials, for example, leads to gassing of aluminium inks; high sulphur content in the mineral oil results in tarnishing of gold bronze inks. Resins with a high acid value tend to a gradual post-wetting of leafing pigments and reduce the flow properties during storage. Therefore the original ink properties may not be maintained during storage.

Most of the printing ink resins used in coloured inks tend to increase viscosity over time in the presence of gold bronze pigments. This is the main reason of the low storage stability of two-component printing inks. In the case of one-component printing inks, good stability in brilliance and viscosity during storage are a must. This requires careful selection of raw materials in the development stage and a permanent quality control of the used resins, ideally in close co-operation with the raw material supplier.

Incorporation of the metallic effect pigments

Metallic effect pigments are very sensitive to shear stress (cf. p. 33 f). Mixing them into the highly viscous binder system of the offset printing ink is obviously difficult. Excessive shear force during mixing can damage the flakes. The resulting fragments and distorted edges of metallic flakes result in reduced brilliance. In fact, metallic pigments do not need to be vigorously dispersed, they only need to be homogeneously distributed in the ink. In practice, the pigments are usually incorporated under vacuum using a butterfly mixer.

One-component offset printing inks are globally available as press-ready products. The MetalStar® range from Eckart is available in various grades for all types of applications such as sheet-fed offset, web offset, waterless offset, low VOC or as UV-curing systems. These products are available in natural colours such as rich gold, rich pale gold and pale gold as well as various silvers and Pantone 871–877 colours.

MetalStar® 07, a mineral oil-based system, has set the world-wide standard for sheet-fed offset inks since its introduction in 1995. The most recent development, MetalStar® SuperGloss, is based on a new generation of even finer metallic effect pigments. This premium product exhibit outstanding transfer, improved yield and perfect half-tone sharpness. At the same time the pigment orientation within the ink film has been improved to lead to high brilliance results. The advantages can be seen in a direct comparison.

Figure IV-2.1.3 emphasizes the fact that with fine, opaque pigments higher colour densities can be achieved - with the same amount (quantity) of transferred ink.

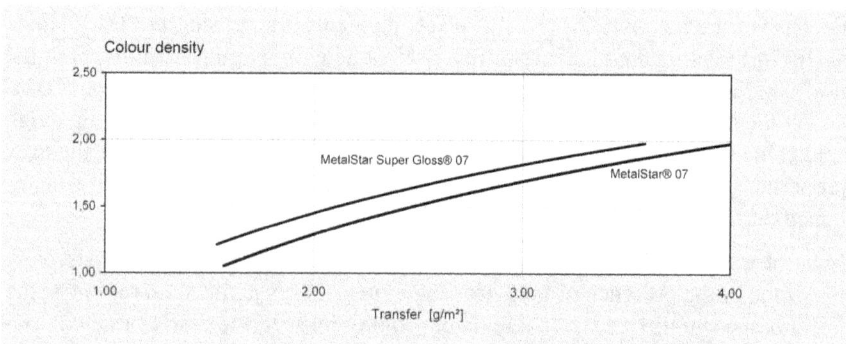

Figure IV-2.1.3: Correlation between colour density and transfer of two different gold inks: MetalStar® Standard uses conventional metallic effect pigments MetalStar® SuperGloss uses extremely fine metallic effect pigments

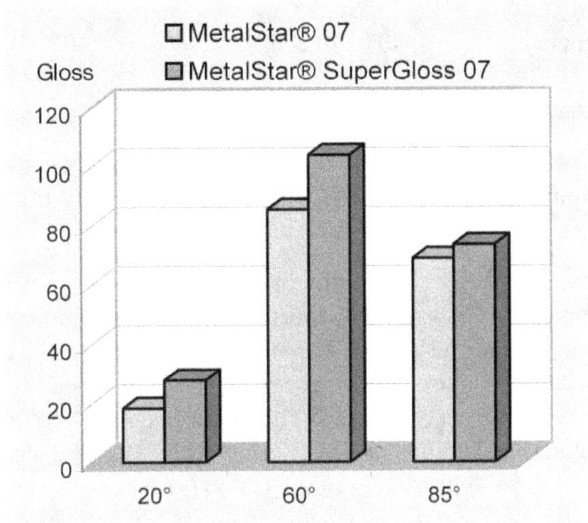

Figure IV-2.1.4: Gloss values of various one component offset inks

Due to the excellent pigment orientation, the new premium inks also exhibit very high gloss. At 20°, 60° and 85° the measured gloss of MetalStar® SuperGloss is above those of MetalStar® Standard Quality (Figure IV-2.1.4).

Mineral oil free offset inks

Offset inks can also be formulated with vegetable oils, which are sustainable raw materials, without any mineral oils. Such systems do not exhibit any physical drying (setting) but cure by oxidation. Besides ecological considerations, these inks offer other practical advantages. As the drying process is entirely based on oxidation these products are highly suitable for non-absorbent substrates such as plastics (these inks are known as "foil inks") or other surfaces with extremely high density. The high vegetable oil content also leads to increased surface tension of the ink film compared to mineral oil-based systems. The surface tension remains almost constant even during the drying process. This provides good flow for subsequently applied layers such as dispersion lacquers or UV coatings. Therefore adhesion on top of these ink systems is good.

UV Systems

Besides printing inks which dry physically or oxidatively, radiation curing systems (UV) are also more and more commonly used. The composition of a UV-ink is, in principle, similar to that of a conventional ink. A typical formulation is:

Pigment powder (gold)	45%
Acrylate oligomer	40%
Photoinitiator	5%
Acrylate monomer	7%
Wax paste	3%

Printing inks pigmented with aluminium contain 15% to 18% effect pigment and accordingly more oligomer/monomer.

Despite the somewhat higher price, the use of UV inks is increasing since they offer the user numerous advantages, most of all instant drying without solvent release. When exposed to UV light, the ink cures completely within 0.1 to 0.2 seconds. The printed product can be processed further immediately. No printing powder is required as no set off occurs. This is for instance important when printing pharmaceutical packaging. In-line coating with UV varnishes does not lead to deterioration of gloss (gloss back) as the UV varnish cannot penetrate the uncured offset ink. Here we also find growing demand for one-component systems. MetalStar® UV products have established themselves as stable, press-ready gold and silver inks for paper, board and various plastics. This range of inks exhibits outstanding transfer, a very good ink/water balance and is therefore ideal for sheet-fed offset. Because of the relatively low film thickness, through-drying problems do not occur.

PVD pigments for offset printing

Besides conventional metallic effect pigments, aluminium pigments manufactured by PVD technology (cf. Part III., Chapter 1.) can also be used in offset printing. These pigments are totally free of lubricants and thus exhibit perfect non-leafing characteristics. Because of this, the corresponding inks are ideal for colouration and enable extremely brilliant, polychromatic effects.

Nowadays, press ready sheet-fed offset inks based on PVD pigments are available. The advantages of these TopStar, inks (supplied by Eckart) are:

- highest brilliance and unsurpassed colour purity
- exceptional metallic effects (mirror finish)
- easy recycling of printed products (free of copper and zinc)
- excellent non-leafing characteristics of the metallic effect pigment
- very good intercoat adhesion
- no loss in brilliance if overcoated

Leafing systems can be obtained by suitable surface modification of the PVD pigment. These inks have obtain brilliance values in offset printing which are almost equal to those of metallised paper.

Guidelines for using offset metallic inks

The substrate

As already mentioned, metallic effect pigments must orientate themselves plane-parallel to the substrate surface in order to achieve brilliant effects. Because of the low film thicknesses in offset printing, larger irregularities in the substrate surface cannot be levelled out. Consequently, paper quality is crucial for a high quality metallic effect. A rough, strongly absorbent surface is impossible to permit excellent print quality. Improvements can only be achieved, for example, by pre-printing a suitable primer. Coated papers are most suitable.

The pH-value of the paper is also important: strongly acidic papers or boards should not be used since acids attack the metallic pigments and can significantly impact print brilliance. In practice, the pH of both sides of the paper should be tested before printing, even if only one side is printed.

The printing press

The printing press should be set up as usual. Modern gold and silver inks can be applied like process inks. Hard rubber blankets have proved to improve the ink transfer but must be carefully levelled. A hard rubber blanket transfers almost all the ink, so that a maximum degree of saturation can be achieved.

The fountain solution

Together with the pH-value of the substrate, also the pH of the fountain solution must be precisely monitored. Ideally it should be in the range of 5.0 to 5.5 for printing gold and silver inks.

Difficulties with metallic offset inks and suggested solutions

The most common difficulties in printing metallic offset inks are described in the following.

Scumming

A metallic printing ink is scumming, if it is partially or completely picked up by the non-printing areas of the printing plate and can no longer be removed with water. This phenomenon is typically related to the following causes:

- the colour density is set too high
- the viscosity of the printing ink is too low
- the impression force of the ink rollers is too high

In most cases, solving this problem is relatively easy:

- apply lower ink densities, where high ink densities are required, the ink has to be applied via two printing units
- add 15 to 20% of a high viscous binder to the ink
- reduce impression force if necessary

Slow drying

An ink is drying too slowly if after 48 hours the ink film still rubs off or the ink show signs of set off within the stack. The problem is particularly critical because it only becomes obvious once printing has finished. Very often the printed sheets can only be used when additionally overprinted with an extra layer of varnish. In general the causes of excessively slow drying are complex. Usually this problem is caused by a combination of several reasons:

- excessive colour density (gold: > 1.6 with Y filter [yellow], silver: > 1.1 with C filter [cyan]) is applied
- the ink is over-emulsified
- the pH-value of the substrate or of the fountain solution is too low

Since the drying problems are only evident after printing, some basic tests should be carried out beforehand in case of printing under new conditions (new machine settings, new unknown substrate, new fountain solution additives). The recommended colour densities of 1.6 for gold (Y filter) and 1.1 for silver (C filter) should not be exceeded.

The amount of fountain solution during printing should be kept as low as possible. Although, because of their different composition, metallic effect inks require a little more fountain solution as coloured inks, over-emulsification must be avoided in any case.
The pH of the paper and fountain solution effect the drying characteristics of the printing ink. Very low values can completely inhibit the drying procedure. Both values must be determined before printing and adjusted to a pH 5.0 to 5.5 if necessary. Substrates which are too acidic need to be replaced.

Set off

Set off occurs, if the ink transfers within the stacked sheets, from the surface of the freshly printed sheet to the back side of the sheet above. There can be several causes:

- the ink dries too slowly
- the pressure in the stack is to high
- the amount of printing powder is insufficient

Set off can be caused by slow drying. Possible solutions have already been discussed above. The problem can however also be solved in a by far simpler way:

- install intermediate shelves in the stack. This not only reduces the pressure in the stack but also enables oxygen to penetrate the stack more easily thus accelerating drying of the ink (in case of oxidative-drying systems)
- increase the amount of printing powder thus reducing the contact area of the sheets in the stack and facilitating penetration of oxygen

Chalking

After drying, the effect pigments can be very easily wiped off. A phenomenon related to metallic effect systems as the particle size of effect pigments prevents their migration into the substrate. If penetration of the binder into the substrate during drying is excessive, the pigments are left without any "glue" to the substrate surface.
The reasons for the problem may be:

- the coating of the paper is too absorbent
- the ink has emulsified
- the pH of the fountain solution is too low

While sealing the surface is helpful in the case of printing on a strongly absorbent substrate, chalking can be visible evidence of slow drying and the solutions discussed above should be used:

- preprint a base coat or coloured ink. If four colours are printed together with metallic effect inks, it is possible in many cases to print half tone yellow underneath the gold or half tone grey under the silver
- regulate the ink/water balance
- adjust the pH of the fountain solution (5 to 5.5)

Recommendations for finishing metallic offset inks

Products printed with metallic effect pigments are often surpassed to a further finishing process in which a varnish is applied to improve gloss (reduction of surface roughness) or to protect the article against abrasion. Laminating with films is also common.

Finishing mineral oil based inks

The most commonly used metallic printing inks are mineral oil-based and nearly all of them are formulated with leafing pigments. The reason is relatively simple: maximum brilliance is achieved with leafing pigments. This enables the printer to compete with other printing techniques. Two factors can cause problems in the finishing process: leafing pigments are not fixed in the ink film, they orientate themselves at the surface of the film which can impact the adhesion of additional layers such as UV varnishes or laminated films. Furthermore, mineral oils in the binder will migrate to the surface of the ink film and reduce its surface tension, thus reducing the wetting of a subsequently applied varnish.

Nevertheless, reality has shown that these mineral oil-based systems can be successfully posttreated if a few simple recommendations are followed during printing:
The thickness of the ink film should be kept as low as possible. This can be achieved by the use of inks with a maximum colour intensity (e.g. SuperGloss®). For stronger metallic ink effects, inks should not be printed as solids, but as a half tones (90%). Ideally, application should be via two printing units. Pre-printing with a coloured ink has also shown advantages.
Analysis of the surface tension of mineral oil-based inks (after application) has shown the ideal time for a finishing process is between 24 to 48 hours after the initial ink application: After 24 hours, the drying of the ink should be complete, after 48 hours a migration of the mineral oil in the ink reduces the surface tension. In-line coating with dispersion lacquers is ideal, subsequent finishing is less problematical.

Any additional surface treatment significantly reduces the metallic brilliance of conventional ink systems.

Finishing of mineral oil free printing inks

Mineral oil free inks provide good overprinting and finishing properties in general. On the one hand, there is a slightly lower leafing effect of the metallic pigments and on the other hand, there is a significantly higher surface tension of the ink film. Both factors contribute positively to the flow of subsequently applied coatings.

Typically these inks dry by oxidation only, therefore it is critical to ensure that drying of the ink film is complete. As discussed previously, film drying depends on individual conditions such as temperature, stack height, type of substrate and pH of the fountain solution. Any subsequent over-coating reduces further migration of oxygen. If the ink layer was not entirely dry, any additional drying is prevented after post processing. This may lead to excessive penetration of the binder and poor bonding of the pigment to the substrate, causing adhesion problems. Apart from that, over-coating also reduces the metallic brilliance of mineral oil-free systems.

Finishing of PVD-based inks

The over-coatability of PVD-pigmented offset inks with their outstanding non-leafing characteristics differs significantly from conventional systems. Because of their perfect non-leafing property, the surface of the ink film is similar to a process colour ink. If the ideal timing between 24 and 48 hours after applying the ink is taken into consideration, there are no differences to regular four-colour process printing to be expected. Over-coating of PVD inks does not impact the brilliance!

Application of metallic offset inks via coating unit

Many offset printing presses are equipped with an additional in-line coating unit, which can be set up to apply water-based metallic effect inks. These inks are available in the MetalStar, range as one-component printing inks. The coating unit inks are basically similar in composition to water-based flexographic inks (cf. p. 170 ff). Compared to offset inks, the low viscous water-based coating unit inks provide more brilliant effects. The pigment particles orientate themselves more easily parallel to the substrate surface and, because of the lower number of ink splits, significantly more ink can be applied (compared to offset units). Since coarser metallic particles can be applied, by far higher levels of brilliance can be achieved.

The metallic ink is applied using engraved cylinder with a chambered doctor blade and metal-backed photopolymer printing plates. This process is most efficient on offset presses with two coating modules. The metallic ink can be printed via the first or second coating unit, depending on the application. On heavily absorbent substrates, a suitable primer can be applied before applying the gold. Alternatively, as with packaging applications, the printed surface is protected by a dispersion lacquer via the second coating unit. Figure IV-2.1.5 illustrates the principles of a double coating unit.

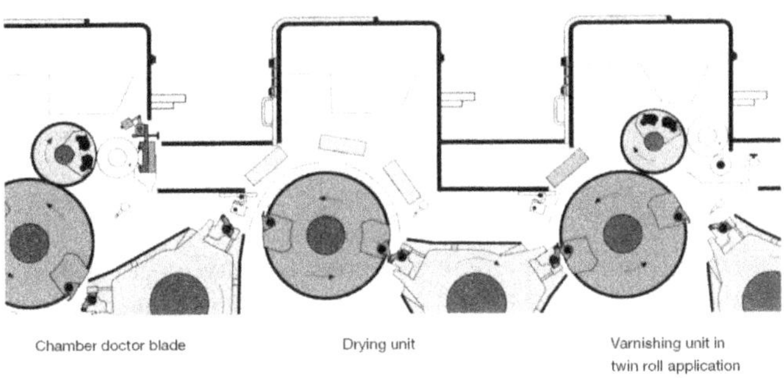

Figure IV-2.1.5: Double coating unit with doctor blade chambers and twin roll application

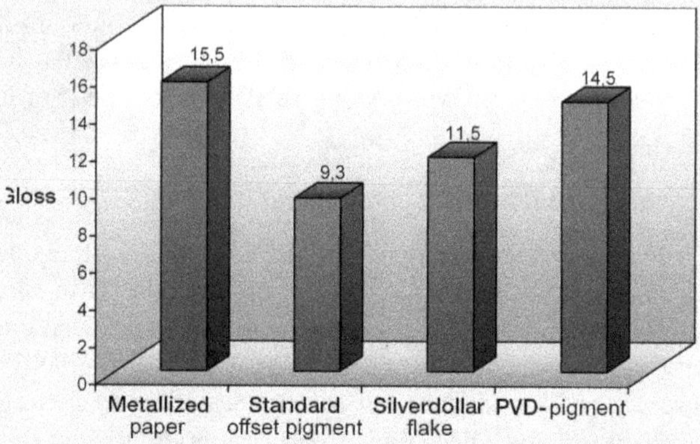

Figure IV-2.1.6: Comparison of the gloss of various types of pigment applied with a coating unit

Besides conventional cornflake grades, PVD pigments and silver dollar pigments (which are primarily used in automotive finishes), can also be used in this printing method. This results in considerably higher levels of metallic brilliance. Figure IV-2.1.6 shows that it is possible to achieve remarkably high brilliance (metallic gloss) with inks based on PVD-pigments, (e.g. TopStar® aqua range from Eckart). This level of brilliance is basically equal to metallised paper.

Such ink systems can be over-coated easily. The best results are again achieved with non-leafing pigments (i.e. silver dollar or PVD pigments). In these cases, the gloss is only effected very little.

References

[1] *Teschner, Helmut:* Offsetdrucktechnik. Fellbach 1989
[2] The printing ink manual. VNR International 1989

2.2 Gravure, flexographic and screen printing

Today's consumer markets are largely saturated and highly competitive. Products are increasingly interchangeable and it is therefore more and more difficult for companies to differentiate them from the massive range on offer. The use of metallic printing inks for decorating products or their packaging offers excellent opportunities to make products stand out and make them more attractive. Gravure, flexographic and screen printing are three common processes for achieving the highest quality metallic effects on printed items.

Printing processes and their commensurate demands on metallic effect pigments

Gravure

Gravure printing is used mainly for high circulation newspapers, mail order catalogues and printed publicity material (e.g. newspaper inserts and brochures). Metallic effects are applied in gravure printing primarily for labels (mainly paper labels), folding boxes (e.g. cigarette boxes), flexible packaging and wall coverings.

Printing process

Characteristic of gravure are steel impression cylinders on which a thin layer of copper has been deposited electrolytically. The printing elements are typically in the form of cells in the surface of the cylinder and are generated by mechanical or laser engraving or chemical etching (Figure IV-2.2.1) [1 to 3].

Figure IV-2.2.1: Gravure printing unit

Individual image areas are separated from each other by very thin ridges lying at the same height. The engraved cylinder is subsequently chrome-plated to increase its durability. Each colour requires a separate cylinder.

Web rotary printing

Nowadays, gravure is largely carried out as a rotary web process. During the printing process, the gravure cylinder is fully inked by rotating in an ink storage tank (Figure IV-2.2.2).

Figure IV-2.2.2: Gravure printing unit: ink storage tank

A steel doctor knife is located shortly before the print area to remove excess ink from the surface. The ridges on the cylinder form the support for the doctor blade and the ink thus remains only in the cells. On non-image areas or ridges, the ink is wiped off and returns to the ink storage tank. In a further step, the paper web is passed against the gravure cylinder using a rubber-covered steel cylinder, and the ink in the screen cells is transferred to the paper. In a drying chamber, directly after the print zone, warm air is used to dry the freshly printed paper web. The printing inks dry physically by evaporation of the solvent. The drying unit is sufficiently long to suit the evaporation rate of the solvent. After drying, the next colour is printed. In contrast to offset printing, the paper web is almost dry before the next print pass takes place (Figure IV-2.2.3).

Areas of application

Because the gravure cylinders are relatively expensive to manufacture and print speeds are high (up to 16m/sec), gravure is generally only viable for high, repetitive print runs (e.g. illustrated magazines). Small print runs and frequently changing motifs are difficult to justify on cost grounds with gravure. Gravure provides the highest quality print with outstanding reproduction.

Use of metallic effect pigments

Before using metallic effect pigments in gravure printing, one must ensure that the particles "fit" in the cells of the gravure cylinder and also easily transfer from these to the substrate. The dimensions of the cells cannot be infinitely large and for detailed copy must be relatively small (the diameter and depth of the cells are in the µm range) thus setting limits on the size of useable metallic effect pigments. Generally, the dimensions of gold pigments do not exceed 10 µm and

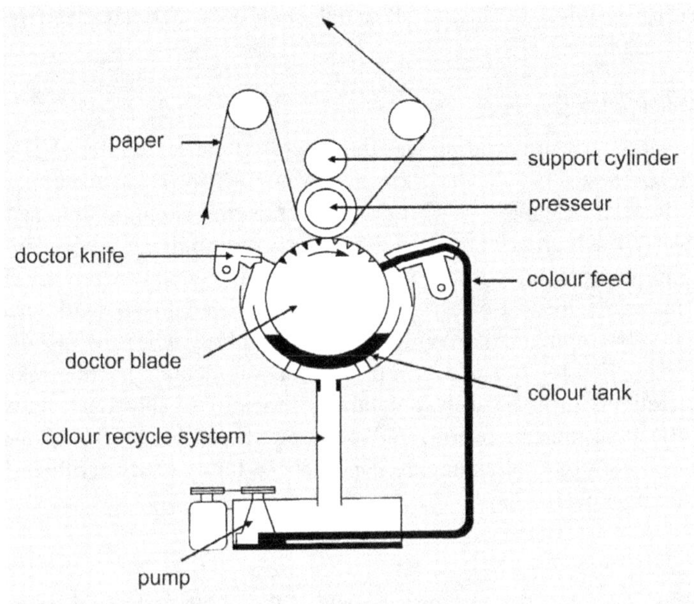

Figure IV-2.2.3 Layout of gravure printing unit [1]

18 µm in the case of aluminium pigments. The rule of thumb, however, is that brilliance increases with the coarseness of the pigment and opacity with fineness (cf. p. 28). The most frequently used pigment grades are in the range 6 to 8 µm in the case of bronze pigments, 8 to 12 µm in the case of conventional cornflake aluminium pigments or 12 to 16 µm in the case of modern silver dollar aluminium pigments (cf. p. 31 f).

Poor print quality can occur with the coarser pigments (build-up on doctor knife, muddy image, etc.), since adequate filling and emptying of the cells and problem-free transfer of the pigment to the substrate are no longer guaranteed. Very detailed copy and strongly absorbent substrates sometimes require even finer pigments such as those used, for example, in offset to ensure problem-free transfer in and out of cells with correspondingly small dimensions. In principle, it is true that compared to flexographic and, even more so, offset printing, the brilliance and metallic effect achieved with gravure are superior because the pigment sizes and ink systems used permit optimal orientation and reflectance of the pigments.

Besides solvent-based systems, low viscosity water-based systems are used in gravure. Typical solvents, particularly in combination with metallic effect pigments, are acetates, alcohols and in some regions (e.g. Asia) toluene, even today, with which the highest brilliance can be obtained. Typical binders are cellulose systems (nitrocellulose, CAB, CAP, ethyl cellulose), acrylates and polyvinyl butyrals.

Use of PVD pigments

For many years, gravure printing was the process for applying the PVD aluminium pigments described earlier (Physical-Vapour-Deposition-Aluminium pigments, cf. Part III., Chapter 1). This dominant position is still maintained primarily because of the high print quality although corresponding systems can be used in offset and flexographic printing (see below). PVD pigment-based printing inks are used primarily as cost-effective alternatives to metallised stock (paper and film) or to foil stamping. Cost savings occur particularly when the metallic effect needs to be applied to only part of a print. PVD-pigmented printing inks can be applied selectively in such cases. In contrast, if metallised substrates are used, the non-metallic areas must be overprinted or even demetallised by an expensive process. Typical areas of application are paper labels, high quality cardboard packaging and flexible packaging.

Flexography

Flexography is a relief printing process and is the oldest industrial printing process. It is used on a major proportion of contemporary paper, board or film packaging and thus competes directly with offset and gravure printing.

Classical relief process

In classical relief printing (letterpress) and flexography, which is more commonly used these days, printing is effected by the raised parts of the printing plate (Figure IV-2.2.4). In flexography, web stock is processed on a rotary printing press equipped with rubber or plastic (e.g. PVC) printing plates [1, 4]. Current print technology is increasingly based on photopolymeric plates, on which the image bearing elements are polymerised, i.e. cured, by incident light irradiation. The areas not exposed to light during production of the plate remain soft and can be removed either by washing out or, in the case of certain plates, blowing out. The finished plastic plate is then glued to a plate cylinder.

Modern flexographic printing

In modern flexographic presses, the low viscosity ink is transferred to a roller covered with tiny ink cells (screen or anilox roller). This can be carried out either using a rubber roller or by dipping the anilox roller directly in the ink reservoir. Excess ink is subsequently scraped off, as in gravure, thus guaranteeing an absolutely uniform ink film. The anilox roller transfers the ink in a uniform layer to the raised areas of the plate from where it is split off on contact with the substrate to which it is transferred.

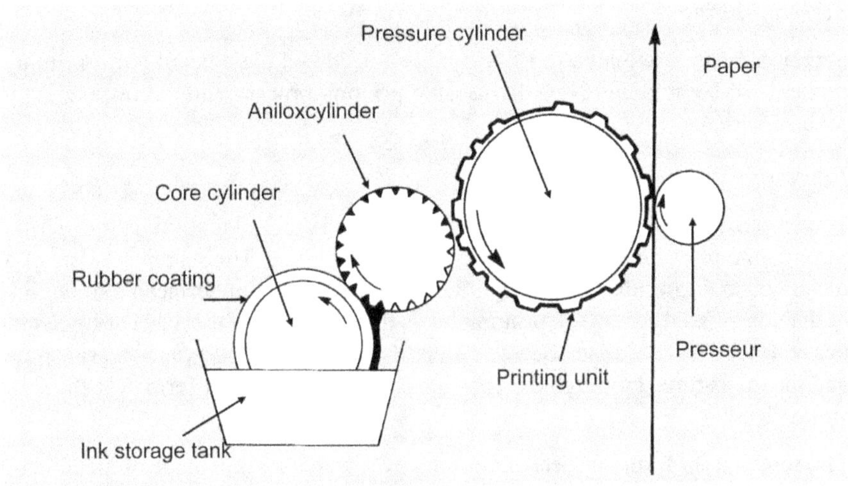

Figure IV-2.2.4: Principle of a flexographic print [1]

Areas of application

Since flexography is suitable for relatively high print runs and can use cheap, particularly thin or rough papers, it is used mainly for printing wrapping paper, film and paper packaging, tissue articles, corrugated cardboard and wallpaper. It is also used for newsprint and advertising media such as carrier bags. Its main advantage over conventional letterpress is the low cost printing plates made of rubber or plastic which are ideal for small runs or frequently changing copy.

Use of metallic effect pigments

In flexography as in gravure, care must be taken to ensure that the metallic effect pigments can easily fill a maximum number of cells on the anilox plate and just as easily exit them again to be transferred to the plate. In contrast to gravure, a further stage, i.e. pigment transfer is necessary. This places additional demands on the mobility and transfer properties of the pigments. Consequently, finer pigments are usually used in flexography than in gravure. Typical particle sizes are 4 to 8 µm (bronze pigments), 6 to 10 µm (cornflake aluminium pigments) and 10 to 14 µm (silver dollar aluminium pigments). Alcohol-based, water-based and UV-curing printing inks of low to medium viscosity are used in flexographic printing. Acetate based inks are generally of only limited use because of the instability of the plate material.

The PVD systems described previously (cf. Part III., Chapter 1) have been recently introduced for flexography and are being increasingly used in this printing process (e.g. for self-adhesive labels, flexible packaging, etc.).

Screen printing

Screen printing originated in China and is considered the oldest known printing process. Today it is a universally useable printing process with correspondingly diverse variants.

Process

Screen printing is based on the principle (Figure IV-2.2.5) that ink is transferred through the print form or screen onto a printing surface. The screen consists of a mesh which allows the ink through the print area and is impermeable to the ink in the non-printed areas. The printing ink is first spread over the mesh using a rubber squeegee. In a second squeegee step, the mesh is pressed onto the substrate and the ink is forced through the open areas of the screen print form onto the substrate [1, 5].

The screen print form consists of a frame, over which a mesh made of metallic or textile fibres or synthetic fibres is stretched. The number of fibres and hence the mesh count can be varied.

All printing stencils are, in principle, the same. The most important point is that the mesh is open on the printing areas and closed on the non-printing areas. Screen printing forms are now produced mainly by photochemical means.

Screen printing machinery

Three types of machine are used for screen printing: flatbed, cylinder and rotary presses. Flatbed presses are equipped with a moveable squeegee and print through the flat screen onto the flat substrate. On cylinder presses, the substrate is fed over a cylinder with a fixed squeegee and a moving print form. In the third type, the rotary screen printing press, the screen printing form is cylindrical and prints on a flat-running paper web.

Areas of application

Screen printing equipment is relatively low cost but highly versatile. It is suitable for printing on the most diverse feedstock, many types of which cannot be printed on at all or only with significantly impaired quality using other methods. Its main use is in such situations which may stem from surface structure or the shape of the substrate or the special requirements of the printing ink. Screen printed products are encountered in every walk of life and include self adhesive films, plastics, paper, cardboard packaging, textiles, metals and all kinds of three dimensional objects. Screen printing is also suitable for short runs.

An advantage of screen printing is the ability to achieve different thicknesses of printing ink by varying the fabric thickness and type of coating. This enables

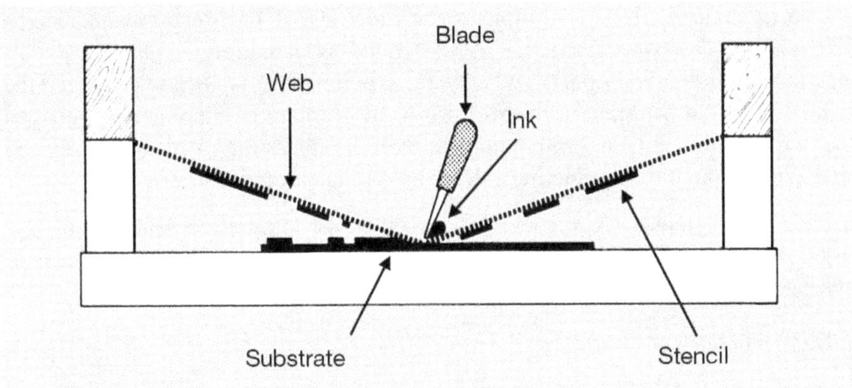

Figure IV-2.2.5: Principle of screen printing [1]

exceptionally glossy print to be obtained which cannot be achieved by other means.

Use of metallic effect pigments

Of all the processes described, screen printing poses the fewest problems on metallic effect pigments because the choice of screen mesh enables a wide range of particle sizes (mean particle sizes up to 50 μm) to be used. The use of relatively coarse metallic effect pigments enables by far the most brilliant metallic and sparkle effects to be obtained with screen printing (cf. p. 25). The ink systems which can be used in screen printing span a wide range of solvent-based, water-based and uv-curing inks of medium to high viscosity.

Properties of metallic effect pigments in gravure, flexographic and screen printing formulations

In contrast to most ink pigments, metallic effect pigments are not chemically inert, i.e. they undergo chemical reactions in various systems, which need to be taken into account when formulating. If they are disregarded, undesirable consequences can result. Excessive mechanical loading can also damage metallic effect pigments (cf. p. 33 f) and result in poor performance.

Chemical reactivity

As already described in detail (cf. p. 36 f), gold-bronze pigments can undergo oxidation leading to the formation of coloured copper salts which is manifested by discolouration (greening or greying) of the metallic effect pigment print. In many cases such reactions are only apparent on the finished printed product after several months, especially if the substrate gives off reactive components in the course of time. A classic example is the greening of bronze pigment inks on PVC papers. The copper and zinc ions liberated by oxidation are also capable of reacting with other components which can, in extreme cases, lead to gelling of the printing ink. An example is the still unsolved problem of the reaction between gold bronze pigments and nitrocellulose with the liberation of nitrous gases and heat accompanied by a sometimes very rapid thickening of the ink.

Aluminium pigments are reducing agents, the best known reaction of which is certainly with water. In this highly exothermic reaction, considerable amounts of hydrogen can be liberated (cf. p. 85 f).

Surface reactions

Besides these purely chemical reactions, physico-chemical surface reactions can cause changes in the lubricant layer which have a marked effect on the pigment properties. Conventional bronze and aluminium pigments are coated during manufacture with a layer of stearic or oleic acid, which determines their wetting characteristics (leafing or non-leafing) and hence their brilliance and hiding power. If, however, the pigment comes into contact with components which possess a high affinity for the pigment surface or can dissolve the lubricant layer, changes in hiding power, stability, and printability can be observed. A classic example is the loss of the leafing effect in the presence of binders with a high acid number or during treatment of leafing metallic effect pigments with organic or inorganic acids such as citric acid, an additive commonly used to obtain non-leafing metallic effect pigments for printing inks.

Stabilisation of pigments

Manufacturers and users of metallic effect pigments have developed a range of stabilisation technologies in recent decades which now enable them to be used in practically all common systems. To meet industry's increasing demands, various aspects must be taken into account when formulating metallic effect printing inks. The most important are examined in the section "Formulation and manufacture of metallic effect gravure, flexographic and screen printing inks".

Mechanical loading on metallic effect pigments

Metallic effect pigments can be damaged by excessive shear stresses such as those which may occur in various places (e.g. on the doctor blade) during printing. This may lead to a change in printability and, possibly, reduced chemical stability. This damage can consist of the surface coating of the pigment (e.g. stearic or oleic acid) being rubbed off with a consequent change in orientation characteristics and, even, in chemical stability. As a rule, this only occurs if components in the printing ink favour detachment or washing off of the coating (see below).

Excessive shear stress on the pigment can however lead to mechanical damage as previously discussed (cf. p. 33 f).

Other properties

Because of their high specific gravity, metallic effect pigments, particularly bronze pigments, tend to settle in printing inks during storage or even during printing. This phenomenon is more pronounced in low viscosity systems and can lead to processing problems unless precautions are taken (e.g. suitable formulation, intensive stirring of the ink, also during printing if necessary, etc.).

Formulation and manufacture of metallic effect gravure, flexographic and screen printing inks

In the manufacture of metallic effect printing inks, care must be taken to ensure complete dispersion of the pigment so that no pigment agglomerates or aggregates remain. Only thus can optimal pigment orientation, brilliance and hiding power be guaranteed. Dispersion should not, however, result in damage to the pigment since this would achieve the opposite effect.

To ensure maximum dispersion, the metallic effect pigment should ideally be gently made into a paste with a certain amount of solvent, water (including a wetting agent) or dissolved binder. Only after having been left to stand for about 30 minutes should the paste be added to the bulk of the binder and any necessary additives. The subsequent dispersion should be carried out with suitable stirrers which permit gentle but complete pigment wetting. Dissolver discs with a diameter approximately half of the internal diameter of the mixing vessel have proved suitable. More aggressive dispersing units such as triple roll or attrition mills result in damage to the pigment and should definitely be avoided. Equipment which operates too gently, such as manually operated apparatus, is also unsuitable since the required level of wetting not achieved.

The necessary dispersion time and speed of the dispersion unit depend largely on the viscosity of the printing ink and the shear energy input and must be determined experimentally for each case. It should be noted that the shear stress on the pigment increases at higher ink viscosity and can, in extreme cases, result in damage to the pigment.

Solvent-based gravure, flexographic and screen printing inks

The formulation of solvent-based metallic inks is largely unproblematic and subject to few conditions. There is still, at the present time, an unsolved problem with the previously mentioned combination of gold bronze pigments and nitrocellulose systems. The various stabilisation methods (e.g. the classic addition of citric acid) still achieve inadequate ink stability and may also lead to a loss in optical quality.

On principle, the following substances should not be used in the formulation of metallic inks since they may adversely affect the quality (brilliance and hiding power) and stability of the inks:

- solvents with excessive water content
- solvents with excessive fat-dissolving power (e.g. acetone)
- binders with a high acid number or a large number of reactive groups
- halogenated binders

Typical formulations contain approx. 30 to 40% bronze or 15 to 20% aluminium pigment and are based on a pigment/binder ratio of around 2:1 or 1:1.

Examples

- Gravure inks

30% bronze pigment		15%	aluminium pigment
15% acrylic resin	or	15%	nitrocellulose
55% isopropyl acetate		5 %	plasticiser
		65%	solvent (ethyl acetate, ethanol)

- Flexographic inks

40% bronze pigment		20%	aluminium pigment
20% PVB	or	20%	nitrocellulose
40% ethanol		5%	plasticiser
		55%	ethanol

Choosing between leafing and non-leafing pigments

The choice in pigments lies between leafing- und non-leafing when formulating for brilliance, hiding power, rub and intercoat adhesion in the final printed product. Intercoat adhesion is the strength of adhesion between two coats of ink or between a coat of ink and a laminated film. The best compromise in properties must be sought for each application. Leafing pigments thus have a high brilliance but moderate rub resistance and intercoat adhesion whereas non-leafing pigments offer rather lower brilliance but in some cases perfect rub resistance and intercoat adhesion.

Use of PVD pigments

Specific formulation parameters must be taken into consideration when using PVD pigments in inks to ensure maximum pigment orientation. Such inks generally contain approx. 3% PVD pigment in gravure and approx. 6% in flexographic formulations. The pigment/binder ratio is a key parameter in achieving various effects. In principle, the achievable brilliance or metallic reflection is greater the smaller the binder content in the formulation.

In counter print on transparent films, where it is possible to achieve mirror-like effects which can be scarcely differentiated from conventional films or metallised films, the pigment-/binder ratio should be adjusted to around 2:1. On applications on paper or board where aspects such as rub resistance are important, the typical pigment/binder ratio is more like 1:2. Depending on overall requirements, this ratio is infinitely variable in either direction to find the necessary compromise between brilliance (mirror effect), hiding power, rub, substrate adhesion and printability. Press ready printing inks with PVD pigments are available from Eckart under the UltraStar® brand. Here all the special formulation parameters have been taken into account so that print quality is ideal.

Water-based gravure, flexographic and screen printing inks

Metallic effect pigments which have been specially stabilised to largely prevent oxidation properties (cf. Part IV., Chapter 1.1.2) are necessary to produce stable water-based metallic inks. The other components (especially binders and wetting agents) in the ink must also be carefully chosen otherwise the quality of the ink or metallic effect can suffer significantly especially in the case of water-based bronze pigment systems. A typical example is provided by systems consisting of bronze pigments and ammonia-neutralised binders (e.g. many acrylates) which tend to suffer a strong loss in brilliance and hiding power and, within a very short time undergo thickening. Binders neutralised with organic amines or inorganic bases are, in contrast, less critical and permit the formulation of stable one-component printing inks. In all cases, however, the tendencies of these and other additives should be checked experimentally to be certain.

Two-component systems

Two-component systems are often used where the required print properties (e.g. stability) cannot be achieved with raw materials which are compatible with metallic effect pigments. This is particularly advantageous if water-based metallic inks are exposed to adverse external conditions, such as elevated temperatures, over a long period.

Use of PVD pigments

For many years, PVD pigments were used exclusively in solvent-based gravure inks because of technical limitations. UltraStar® aqua water-based gravure, flexographic and screen printing formulations have been available, however, from Eckart for about two years.

UV-curing gravure, flexographic and screen printing inks

Just as with water-based metallic inks, certain aspects must also be considered with UV inks in order to formulate attractive, and, in particular, stable systems.

This is especially important as otherwise premature polymerisation can occur. The suitability of each ingredient must be checked experimentally. It cannot be excluded that high quality grades (e.g. of mono- or oligomers) with only slight levels of contamination will need to be used to achieve the required ink stability. Particular care should be taken when using PVD pigments. Since the stabilisation of such pigments is relatively complex, press ready stabilised inks are supplied by Eckart under the UltraStar® UV name.

Processing metallic effect gravure, flexographic and screen printing inks

In printing using the processes discussed here, as in all applications of metallic effect pigments, the pigment must be oriented as parallel as possible to the substrate to ensure maximum brilliance and hiding power. The process parameters (press settings, choice of ink pumps, etc) should be chosen to preclude changes in the ink. With existing ink formulations and optimised machine settings, high quality results are certain if high quality stock is used and the amount of ink applied and drying time guarantee efficient pigment orientation. In practice, such optimised conditions are often difficult to achieve because of external constraints or can only be determined for an existing ink system by extensive testing. Various aspects which can be taken into preliminary consideration in order to avoid undesirable effects are listed below although no claim is made as to completeness.

The maximum metallic effect can only be achieved on high quality (coated) papers. Absorbent papers frequently permit only moderate effects. In such cases, the brilliance and hiding power of the metallic ink can be improved by preprinting with a primer. Water-based metallic effect inks, in common with all water-based coatings, tend to foam. However, the lamellar structure of metallic effect pigments makes defoaming significantly more difficult. This should be considered when choosing ink pumps (peristaltic or gear pumps are, for example, preferable).

As mentioned earlier, metallic effect pigments can be damaged by excessive shear stresses. This can occur even during printing if high mechanical forces in the line act on the pigment, for example as a result of incorrect doctor blade settings.

References

[1] *Leach, R. H., Pierce, R. J.:* The Printing Ink Manual, 5th edition, London 1993
[2] *Dohmen, W.:* Der Tiefdruck, 3rd edition, Köln 1986
[3] *Ollech, B.:* Tiefdruck, 2nd edition, Bielefeld 1993
[4] DFTA: Technik des Flexodrucks, 3rd edition, Stuttgart 1991
[5] *Ehlers, K. F.:* Siebdruck, 3rd edition, München 1980

3 Plastics

Applications and processing

As it is the case with all other effect pigments, following points have to be particularly noted when processing metallic effect pigment flakes:

- transparency of the polymer matrix: it is very difficult to produce a metallic effect with opaque plastics
- optimal dispersion: the pigment particles must be completely and homogeneously dispersed in the polymer without damaging the flakes by too vigorous a dispersion
- the metallic effect pigments must be oriented as parallel as possible in the plastic

Whenever processing plastics containing metallic effect pigments by injection moulding or extrusion, an intermediate step – the production of a masterbatch – is usually necessary, even if the process involves colouration with only one pigment. The reason for this is the small quantity of pigment used and the relatively difficult handling of metallic effect pigments. Up to 40% of aluminium or 60% of gold-bronze pigment should be incorporated into a highly compatible carrier polymer, the exact composition depends on the final application of this masterbatch. The resulting plastic goods however rarely contain more than 3% of metallic effect pigment – with fine pigments this figure is usually lower, typically less than 0.5%.

Processing metallic pigments compared with colour pigments

As already mentioned earlier, the metallic effect can only come up properly, if the flake shaped pigments are optimally wetted and oriented parallel to the surface (cf. p. 33 f). To obtain an optimum of effect, the pigment particles must be completely dispersed and distributed so that no agglomerates, with large sized particles, are visible to disturb the optical effect.

Of course, the very thin, flake-shaped particles are very easy to bend or break during processing, which may eventually lead to greying (cf. p. 33 f). It is not easy to determine the correct amount of shear force to apply during processing without destroying the pigment. Effect pigments, produced by the pigment manufacturer in the form of granules, are an improvement. The pigment particles are pre-dispersed in a matrix and do actually not agglomerate further, so that a high shear force is no longer needed. Figure IV-3.1 shows pigment particles after exposure to low and high shear forces.

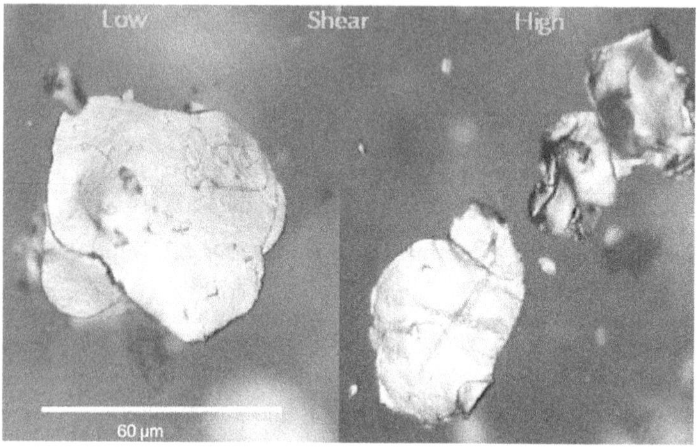

Figure IV-3.1: Effect of shear force on metallic effect pigments ("Cornflake type")

Furthermore, use of opaque pigments should be avoided as far as possible, as they affect or even virtually destroy the light reflecting effect of the metallic gloss. Therefore, for colouring of the polymer, only transparent colour pigments should be used. Most organic pigments meet this requirement. For inorganic pigments however, the choice is very limited and only ultramarine blue and violet and transparent iron oxides have any major importance. If a metallic effect needs to be brightened up, a white (transparent) pearlescent pigment should be used instead of TiO_2 which is strongly opacifying. The same principle applies for the selection of the polymer: the more transparent the polymer, the more brilliant is the metallic effect achieved. Most fillers and reinforcements are strongly opacifying and should therefore only be used as little as possible.

Metallic effect pigments for plastics

Aluminium pigments

Conventional non-coated aluminium pigments are suitable for colouring almost all plastics. They are stable up to a temperature of over 300 °C and have very good light and weathering stability. Depending on the desired effect, cornflake pigments or the very bright round silver dollar pigments are used (cf. p. 315). Only with very aggressive plastics such as PVC, problems with processing can occur. At high processing temperatures and high shear forces, insufficiently stabilised PVC can release hydrogen chloride. This can chemically react with the aluminium pigment resulting in immediate black discoloration. In such cases coated aluminium pigments, resistant to hydrogen chloride, e.g., Aluminium Resist,-grades (cf. p. 35 f) (Table IV-3.1) must be used.

Because of their very high resistance, coated effect pigments can produce extremely good weathering results with guaranteed long-term durability.

Coloured aluminium pigments

Coated, coloured aluminium pigments are a speciality (cf. Part III, Chapter 2). They are currently available for plastics in many colours, e.g., gold, blue and green. After surface coating with colour pigment, the metallic appearance of the aluminium pigment is slightly reduced but remains clearly visible (Mastercolor® by Eckart). The colour pigment on the aluminium surface is encapsulated by a coating procedure, creating very good stability for plastics applications. The heat resistance is more than 300 °C and the light fastness and weathering stability are outstanding. These pigments are intended for use in two-tone systems. By addition of a second colour pigment (usually darker than the coloured aluminium pigment) in the polymer, a combination of two individual colours results. Visually these two colours remain separate as the coloured aluminium pigment has a large particle diameter. For example, a gold sparkle can be seen in a blue background without the yellow and blue merging into a green.

Gold bronze pigments

Non-coated gold-bronze pigments used in printing and coatings have only very limited suitability for plastics as these pigments have a maximum heat resistance of only 180 °C. Only LDPE-films and various PS- und SB-injection moulded articles can be coloured with them. For all other applications such as colouring PVC, coated (and therefore more resistant) pigments, such as Eckart's Goldbronze Resist®, must be used. These have a heat resistance up to about 260 °C to 280 °C. The light fastness is very good but the weathering resistance is poor. The addition of suitable light stabilisers can raise the weathering resistance to an acceptable level. In Figure IV-3.2 the heat resistance of coated Goldbronze Resist, is compared with that of an uncoated pigment. The test was done according DIN EN 12877 in PP.

Table IV-3.2 gives an overview of the gold-bronze pigments used in plastics.

Forms of supply

Aluminium pigments

Because of the danger of dust explosion, aluminium pigments are practically no longer used in powder form in Central Europe. However in the past, powder was very popular as it was suitable for all plastics and had a price advantage over pastes and granules (pellet) pigment preparations. Pigment pastes are available with different plasticisers and oils. The most common for plastics other than PVC is

Pigment type	Product example	Delivery form	Particle size	Opacity	Brilliance	Effect
uncoated aluminium pigment	STAPA® Reflexal VIII	powder and paste	10 µm	very good	medium	homogeneous silver
uncoated aluminium pigment	STANDART® Reflexal 190	powder and paste	35 µm	good	good	lightly sparkling
uncoated aluminium pigment	STANDART® Reflexal 210	powder and paste	75 µm	medium	very good	strongly sparkling
uncoated aluminium pigment	STANDART® Reflexal 135	powder and paste	235 µm	low	very good	heavily sparkling
uncoated aluminium pigment	Mastersafe® silver 10203	pellets	10 µm	very good	medium	homogeneous silver
uncoated aluminium pigment	Mastersafe® silver 35203	pellets	35 µm	good	good	lightly sparkling
uncoated aluminium pigment	Mastersafe® silver 75203	pellets	75 µm	medium	very good	strongly sparkling
uncoated aluminium pigment	Mastersafe® silver 235203	pellets	235 µm	low	very good	heavily sparkling
coated aluminium pigment	STAPA® Resist 801	powder and paste	18 µm	very good	medium	homogeneous silver
coated aluminium pigment	STAPA® Resist 214	powder and paste	39 µm	good	good	lightly sparkling
coated aluminium pigment	STAPA® Resist 211	powder and paste	60 µm	medium	very good	strongly sparkling

Table IV- 3.1: Overview of aluminium pigment grades used in plastics

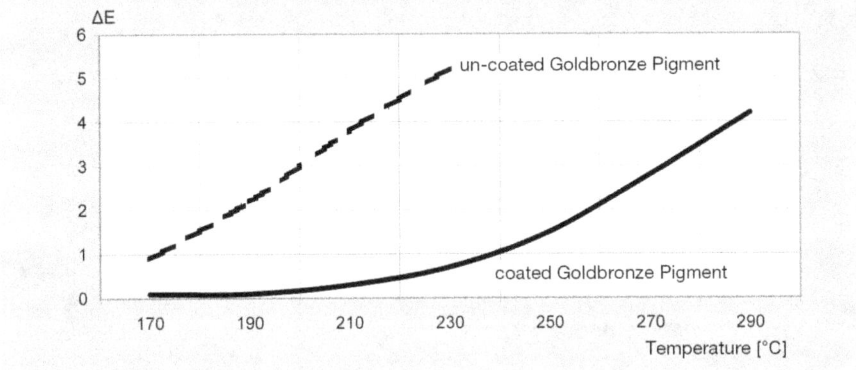

Figure IV-3.2: Comparison of heat resistance of un coated and coated Goldbronze Pigment STANDART RESIST®

paraffin oil or medical grade white oil. For colouring PVC the whole range of PVC plasticizers such as diisononyl phthalate DINP or diisooctyl adipate DOA is available. Because it is genetically harmful, diisooctyl phthalate DOP is used less and less and is substituted by the two other plasticisers already named. Commonly the pigment concentration in commercial pigment pastes is 80%. This guarantees dustless processing while the amount of paste forming material is still not too high so that a highly filled masterbatch is possible.

As an example, to produce a 40% LDPE-masterbatch, the 80% pigment paste is mixed 1:1 with LDPE and extruded. A masterbatch is obtained which has 50% LDPE and 40% pigment with only 10% oil.

On the world market, however, 65% pigment pastes are still available. Starting from there, 40% masterbatches are produced. These contain over 20% additive and a significantly reduced amount of binder. Such a masterbatch is difficult to produce, store or process. A certain disadvantage of pastes is that they do not have the same versatility as powders. That is to say that if, in addition to PVC, it is desired to colour another plastic, e.g., PP, two different pigment pastes will be needed.

Pellets are especially popular in colouring plastics as, in addition to better handling and a much better shelf life, they are almost universally applicable. Only when used in paste-like plastics, such as PVC-plastisol, there are problems with dispersion. In addition to dispersing agents and binders, these pellets usually contain 80% aluminium pigment.

Pigment type	Product example	Delivery form	Particle size	Opacity	Brilliance	Effect
uncoated gold bronze pigment	STANDART® 9900	powder and paste	6 µm	very good	medium	homogeneous gold colour
uncoated gold bronze pigment	STANDART® G 900	powder and paste	28 µm	good	good	lightly sparkling
uncoated gold bronze pigment	STANDART® OT	powder and paste	75 µm	low	very good	heavily sparkling
coated gold bronze pigment	STANDART® Resist Roroflex	powder and paste	8 µm	very good	medium	homogeneous gold colour
coated gold bronze pigment	STANDART® Resist CT	powder and paste	28 µm	good	good	lightly sparkling
coated gold bronze pigment	STANDART® Resist OT	powder and paste	75 µm	low	very good	heavily sparkling
coated gold bronze pigment	Mastersafe® gold 06103	pellets	6 µm	very good	medium	homogeneous gold colour
coated gold bronze pigment	Mastersafe® gold 17103	pellets	17 µm	good	good	lightly sparkling
coated gold bronze pigment	Mastersafe® gold 35103	pellets	35 µm	medium	very good	heavily sparkling

Table IV-3.2: Overview of the gold-bronze pigment grades used in plastics

Figure IV-3.3: Aluminium pellets

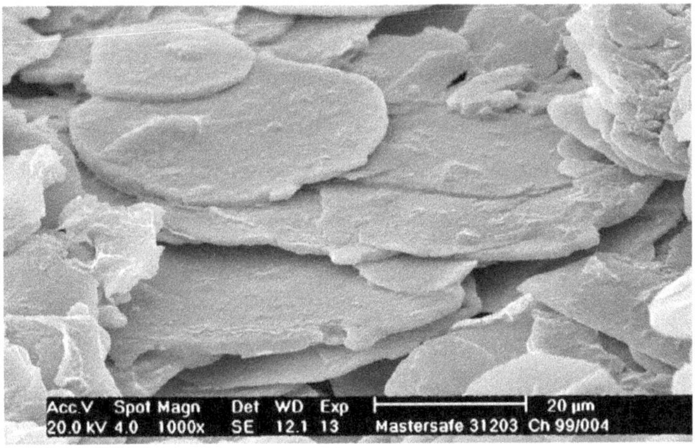

Figure IV-3.4: Aluminium flakes being present in pellets

In Figures IV-3.3 and IV-3.4 such a pellet is shown at various magnifications. It can be clearly seen that the pigment flakes are fixed to each other by a small amount of binder without being completely surrounded by binder. This makes it easy to incorporate them homogeneously into the polymer.

Gold bronze pigments

Up to now, most gold bronze pigments are used in powder form as there is no danger of dust explosion and the density of these pigments is so high that there are practically no dust problems.

Same as with aluminium pigments, gold bronze pigment pastes can be formulated with all common plasticisers and oils. These concentrates are commercially available. Since 2001, pelletised gold bronze pigments have also been available for plastics. Because of the very high pigment loading (90%) and very good predispersion of the pigment particles in the binder, no change is usually necessary in the formulation compared with powder pigments. They can be replaced on a 1:1 basis.

These pellets also have very good storage stability without problems of agglomeration (Figure IV-3.5).

Figure IV-3.5: Goldbronze pellets

Orientation problems on injection moulding

The flake shape of metallic effect pigments causes visible orientation along the flow direction of the matrix. In extrusion this can produce orientation similar to that of a coating and thus lead to an optical flop effect. In case of injection moulding applications, flow lines (joint lines) cannot be avoided and are very clearly visible. Opposite to metallic pigments, organic and inorganic colour pigments have an isometric form and therefore the flow lines are not visible. Figure IV-3.6 illustrates the different orientation of the pigments (here: aluminium pigments) along the flow line. On the left side of the picture, the flowing material is shown with effect pigments oriented parallel to the surface – not only on the surface, but also inside the front area. The right picture shows that after the flow fronts meet, the (now incorrect) pigment orientation remains. The surface of the pigments appears bright and the edges remain dark, so that the flow line is clearly marked. Figure IV-3.7 shows how the flow line marks arising from particle orientation; the specific orientation of the pigment particles remains visible.

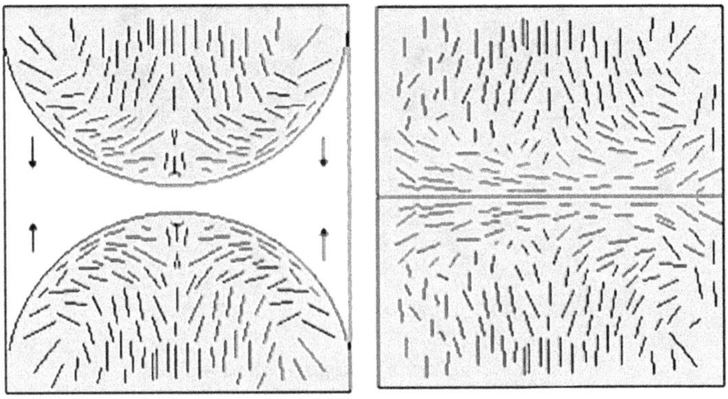

Figure IV-3.6: Formation of flow line
Left: flowing material with effect pigments which orient themselves parallel to the surface
Right: after meeting of the flow fronts the now incorrect orientation remains. (Photo: Basell)

The dwell time as a quasi-liquid state is very short in plastics processing, immediately after that, the flake shaped pigments are frozen into the flow direction. When looking at the structure of the weld line, the surfaces of the pigments appear bright and the edges dark. This light-dark difference pronounces the weld lines. Similar effects can be seen with other flake shaped pigments, for example, pearlescent pigments. The origin always lies in the anisotropy of the effect pigment.

Reducing the visibility of the weld lines

The visibility of the weld lines can be reduced by the following methods:

- Whenever possible, optimal wall thickness and injection point positioning should be considered at the design stage. In many cases this allows optical defects to be hidden away in non-visible areas.
- Flow lines and welds are greatly reduced by using the largest possible pigment diameter and highest possible pigment concentration. However, it has to be mentioned here, that a higher pigment loading can have a negative impact on the mechanical stability of an injection moulded article. With PMMA, for example, the stability can be reduced by approx. 30%.
- Use of darker colours also reduces the optical impact of weld lines. With mixtures containing carbon black or dark blues or greens combined with large pigment particles, even pronounced weld lines are absorbed into the background.

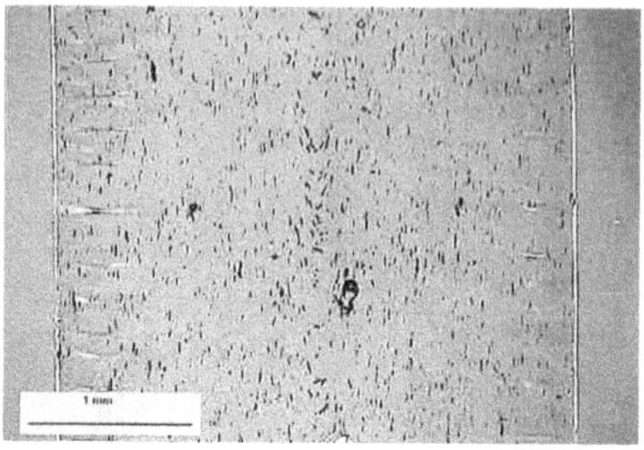

Figure IV-3.7:
Thin layer of an injection moulded polypropylene item with aluminium flakes

- Generally, the manufacturing technique can influence the optical impact of the weld lines. The most satisfactory type of gate is a film gate. A sprue gate is preferable to a pin gate. A cone gate is particularly advantageous with tube-shaped injection moulded articles. A larger gate cross section and higher injection velocity create turbulences in the moulding and this reduces the visibility of the flow lines and welds.
- Polymers with a viscosity as high as possible lead to less orientation of the pigments and therefore to reduced visibility of the flow lines and welds.
- Cascade gates can be used for large items with long flow paths.
- The In-Mould Lamination Process of extruded, silver coloured thermoplastic sheets meets the highest demands. This extrusion method produces optimally oriented metal flakes. If opaque substrates are used, they can be in-mould laminated as much as required.
- For particularly complex moulded parts and high volume production, the SCORIM technique combined with In-Mould-Heating can produce objects free from optical defects (see below).In this case the extra coating step of injection moulded items can be avoided.

A combination of the above methods usually allows moulded articles with metallic pigments to be produced even in difficult cases.

The SCORIM process in combination with In-Mould-Heating

SCORIM = Shear Controlled Orientation in Injection Moulding (developed by the British Technology Group, marketed by the firm Cinpres Ltd. [1])

During mould filling, the plastic melt can be manipulated so that a partial re-orientation occurs in the area of the weld lines. This also leads to significantly increased mechanical stability of the plastic. At the same time the weld line becomes much less obvious. However this can only be achieved with increased flowability of the melt, that means working at a higher melt temperature. Additionally, the mould is heated exactly at the area where the material flows together. The weld line integrates into the surface structure of the plastic. Combination of the SCORIM process and In Mould Heating enables the finish to be greatly improved. A remarkable extension of the cycle time and significantly high costs of purchasing the special equipment considerably increase the unit cost. However, the avoidance of an extra coating step, which is normally necessary with conventional methods, can overcompensate for this.

The SCORIM process in detail

The SCORIM head (Figure IV-3.8) is fitted between the cylinder head and the mould. It consists of a melt separator (here: two gates) and two pistons which pump the melt backwards and forwards into the mould.

Figures IV-3.9 to IV-3.13 show the details of the SCORIM process. The injection points (at least 2) must be positioned so that all possible weld lines lie between them. This technique allows considerable improvements in the surface quality of injection moulded articles coloured with flake shaped effect pigments.

In the first step (Figure IV-3.9) the polymer is injected into the mould as usual. The mould is only filled up to about 90%. On the left section of the picture the appearance of the weld line is shown. The arrows indicate the pumping direction of the pistons which force the material through the mould to the other piston.

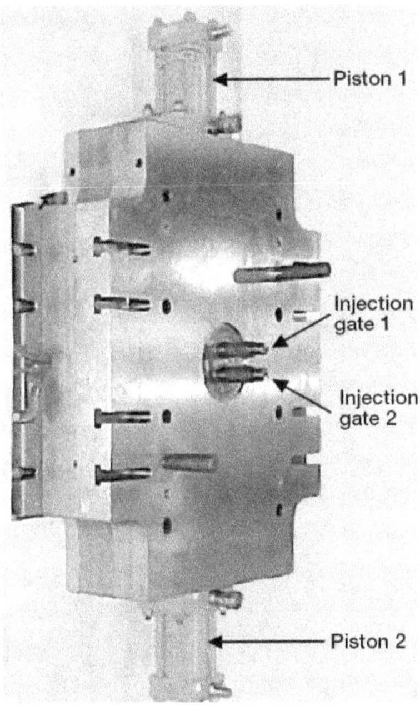

Figure IV-3.8: SCORIM head

In the second step (Figures IV-3.10 to IV-3.13) the melt is hydraulically moved backwards and forwards in the mould. The extra 10% of the material is supplied via the first piston. Depending on requirements, this process can be repeated two or three times. As shown in the graph, the weld line becomes increasingly broader and this "smudging" means that it is usually no longer visible.

In a third step, both pistons compress the melt (as in the hold pressure) until it has cooled in the mould. This compensates for the shrinkage which always occurs.

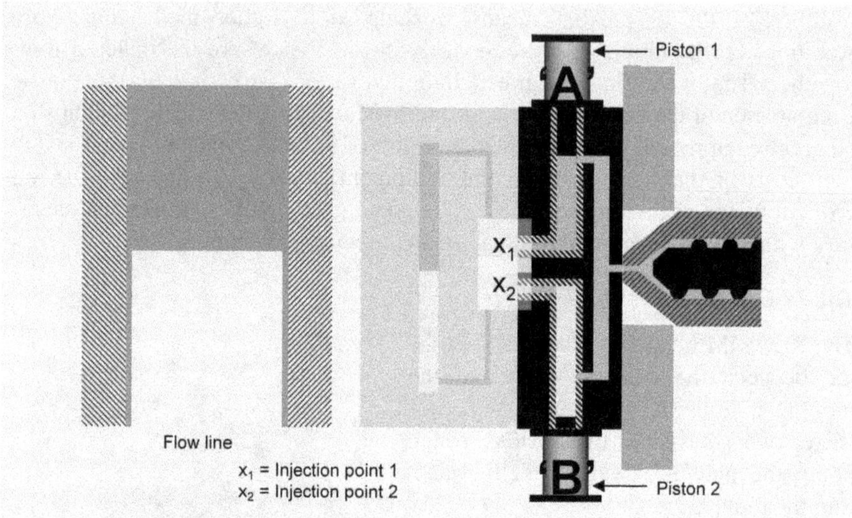

Figure IV-3.9 First step: injection of hot polymer compound

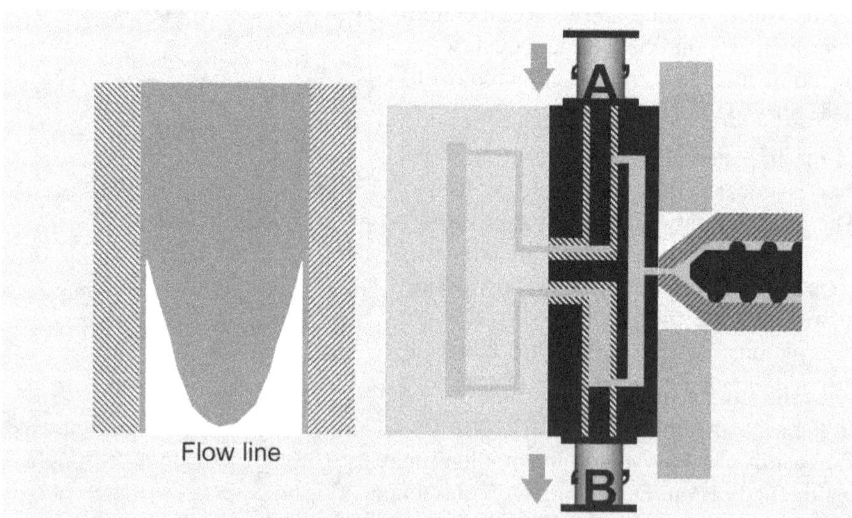

Figure IV-3.10: Second step: first pumping step

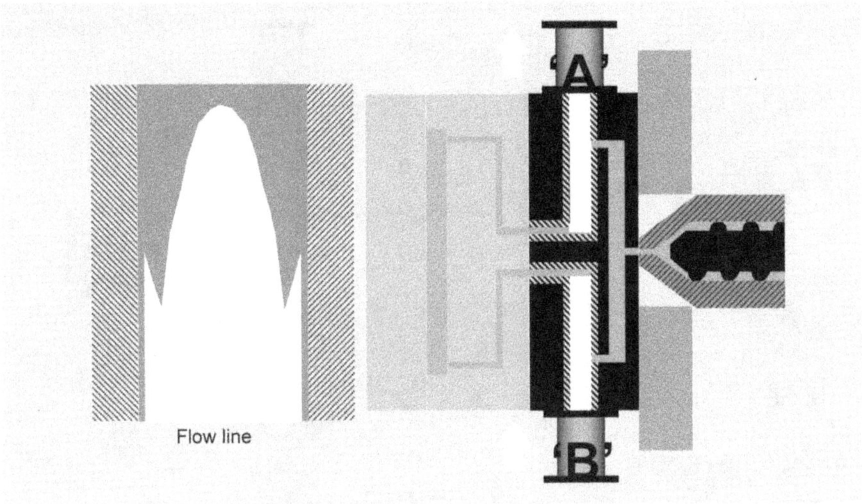

Figure IV-3.11: Second step: second pumping step

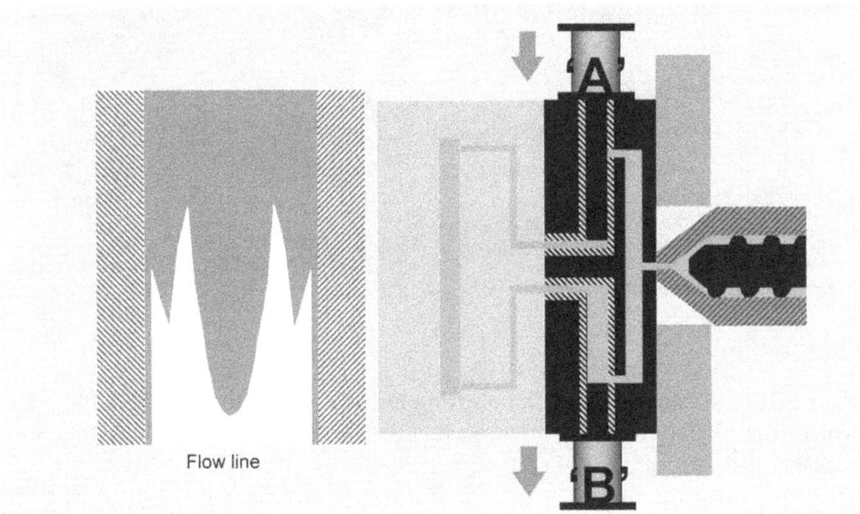

Figure IV-3.12: Second step: third pumping step

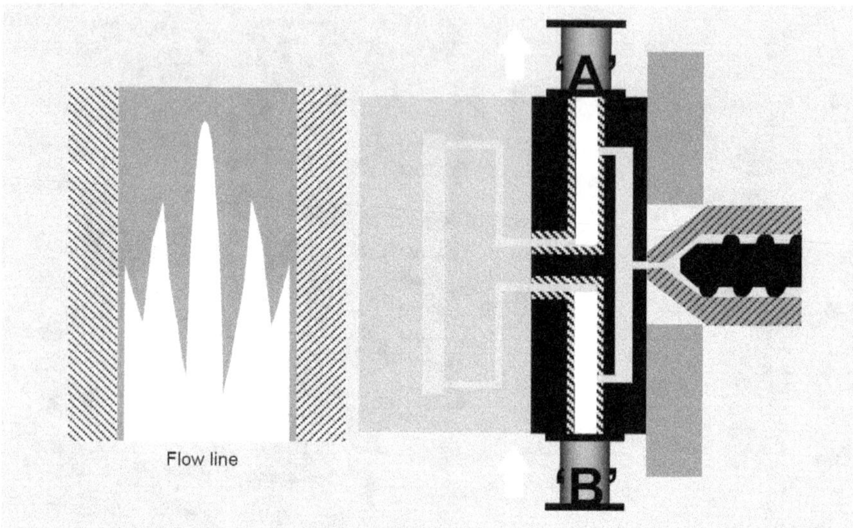

Figure IV-3.13: Second step: fourth pumping step

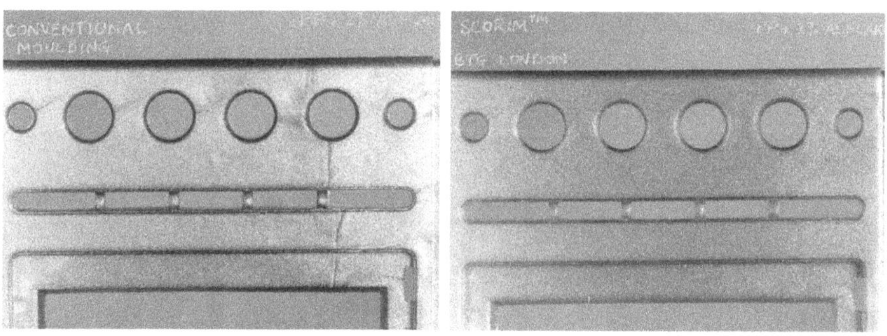

Figure IV-3.14: Weld lines with and without SCORIM

Figure IV-3.14 shows the results of using the SCORIM process. The weld lines which occur when the melts flow together are so smudged in the SCORIM process that they are practically no longer visible.

References

[1] Cinpres Ltd., Ninian Park, Ninian Way, Tamworth, Staffordshire, B77 5ES, UK, http://www.cinpres.de

4 Technical applications

In previous chapters the emphasis has been on the optical properties of metallic effect pigments. In this section applications will be brought together in which metallic effect pigments react chemically. These mainly make use of the high energy content and high reactivity of flakes. In contrast to all previous chapters the term "pigment" will not be used in this chapter. By definition a pigment is inert and does not dissolve in its matrix and this is not true of metal flakes in technical applications.

Properties and requirements

The metallic flakes must have the requisite reactivity for the particular technical application. This reactivity can be obtained by using small particles with a large surface area. The required large surface areas can be obtained by the milling process fully described earlier (cf. Part II, Chapter 2).

To produce these products as economically as possible, the metal must be available in sufficient quantity and be easily workable. In practice it is mainly aluminium-, magnesium- und zinc flakes which are used. In spite of their high reactivity, the flakes must be simple to process. The reaction products produced must not be toxic and must not disrupt the chemical process.

Properties

The reactions of flake shaped metals can be basically divided into three main groups:

- reactions in which metal flakes are sources of energy
- reactions of metal flakes as gas-forming agents
- reactions of metal flakes as reducing agents

Reactions of metal flakes as sources of energy

This makes use of the high affinity of reactive metals for oxygen. The large surface area of the metal flakes means that in the presence of oxygen the activation energy is overcome and spontaneous oxidation occurs. The liberation of thermal energy is controlled by the availability of oxygen (Figure IV-4.1).

The high affinity for oxygen and the large surface area of the flakes are also the origin of the risk of a dust explosion. If the metal forms a dust in air, only a small ignition energy is needed to set off an explosion.

$$2\,Al + 1{,}5\,O_2 \longrightarrow Al_2O_3 + 1677\,kJ$$
$$Mg + \tfrac{1}{2}\,O_2 \longrightarrow MgO + 602{,}1\,kJ$$
$$Zn + \tfrac{1}{2}\,O_2 \longrightarrow ZnO + 348{,}5\,kJ$$

Figure IV-4.1 Oxidation of selected metals

Reactions of metal flakes as gas-forming agents

Some reactive metals, e.g., aluminium and zinc, dissolve in acids or bases liberating hydrogen. This property is made use of in various technical processes. In these processes the time profile of hydrogen formation is of decisive importance.

Metal	Molar mass [g/mol]	Density of metal [kg/dm^3]	Hydroxide	Vol (H_2)* [dm^3/g]
Aluminium	26.98	2.70	$Al(OH)_3$	1.25
Magnesium	24.30	1.74	$Mg(OH)_2$	(0.92)
Zinc	65.39	7.14	$Zn(OH)_2$	0.34
* Volume of hydrogen under standard conditions liberated by 1 g Metal				

Table IV-4.1 Comparison of liberation of hydrogen by different metals

In Table IV-4.1 it can be seen that, mass for mass, aluminium produces the most hydrogen. In addition to aluminium, zinc can also produce hydrogen. Aluminate ($Al(OH)_4$-) and zincate ($Zn(OH)_4^{2-}$) ions go into solution. In contrast magnesium precipitates out as the hydroxide and, under basic conditions, a protective layer of magnesium hydroxide forms on the metal surface, which inhibits further oxidation.

In strong acids, the three metals all dissolve producing hydrogen. In oxidising acids (e.g. nitric acid) aluminium forms an oxide layer which inhibits complete oxidation of the metal.

Reactions of metal flakes as reducing agents

The ability of reactive metals to give up electrons easily is used in many chemical processes (Table IV-4.2). In order to obtain a fast reaction and rapid dispersion of the metal as a reducing agent, flakes are frequently used.

Use of magnesium in basic conditions as a reducing agent comes up against the same problem as when using it as a gas forming agent: the metal surface becomes covered with a layer of magnesium hydroxide which passivates the flakes and inhibits further reaction.

To be able to produce metal flakes, the metal must be ductile enabling flakes with a large surface area to be produced. Large surface areas make possible fast,

Metal	Standard potential	
	in acid solution (pH = 0)	in alkaline solution (pH = 14)
Aluminium	−1.662 V	−2.330 V
Magnesium	− 2.363 V	−2.690 V
Zinc	− 0.763 V	−1.215 V

Table IV-4.2: Standard potentials of some metals

defined reactions. They take place at the boundary between the metal and the gas or liquid. In practice, for applications in which large surface areas are needed, aluminium flakes are preferred over zinc or magnesium flakes, not least because of their ease of handling and availability. In general, to alter the initial stages of the reaction, the surface coating of the flakes is modified. For reactions in aqueous media the wettability plays an important role. The faster the reactants can reach the metal surface the quicker the reaction will start. Common methods of affecting the start of reaction are coating the surface with fatty acids, defined formation of an oxide layer and inhibition of the surface by chemical substances. The duration of reaction increases with increasing particle size.

Aluminium flakes are available as powder and as pastes with organic solvents or water. The use of powder or paste with a particular solvent is determined by the field of application and the technique used. The most important disadvantage of powders is the danger of a dust explosion if handled incorrectly.

Test criteria for metal flakes

The most important tests are considered below.

Metal content

The metal content [1] indicates the amount of the flakes which is present in metallic form. As it is only this proportion which will take part in the various reactions, it determines the quantity of flakes which will need to be used. The metal content is determined by completely oxidising the product in a strong base or strong acid. The hydrogen liberated is measured volumetrically and after correcting to standard conditions the metal content is calculated.

Particle size distribution

One of the most important tests on metal flakes is determination of the particle size. As already mentioned, the duration of reaction and therefore the suitability for a particular field of application depends above all on the particle size. For flake shaped metals this is most effectively determined by limiting particle sieving. This is usually carried out by wet sieving in a suitable solvent [2], as dry sieving would involve the product being whirled around with the danger of a dust explosion. Particle size distribution is generally determined by laser granulometry but this delivers only a relative value as the mathematical model used in analysing the data is based on assuming spherical rather than flake shaped particles.

Surface area determination

The "free surface" [3] is measured by a method well established in the cement industry using a Blaine Apparatus standardised to DIN 66127 (corresponds ISO 10070). A sample is packed to a specified porosity and a gas, normally air, is passed through. From the resistance to the flow, the size of the surface is calculated using the Carman-Kozeny equation [4].

Applications in the building materials industry

The ability of aluminium to react to produce hydrogen under basic conditions (as described above) is exploited in several branches of the building industry and the main applications will now be considered in more detail.

Metal flakes for the manufacture of autoclaved aerated concrete (AAC)

Autoclaved aerated concrete is a modern building material. Low densities with relatively high compressive strength, good heat insulating properties and easy machining are made possible by a pore content over 80%.

Manufacture

The manufacture of autoclaved aerated concrete is shown schematically in Figure IV-4.2. Finely ground quartz-containing sand, cement, lime and water are homogenised and in a final mixing step treated with a pore-forming agent such as aluminium flakes. In some cases, fly ash from combustion plants is used in autoclaved aerated concrete manufacture because of its high silicic acid content. After pouring into moulds expansion begins. This involves increasing the volume of the cement by formation of finely distributed pores. After the end of the expansion and reaching the required green hardness (the solid state achieved before steam curing), cutting takes place. The following hydrothermal process, which involves treating the green material (that is the unhardened autoclaved aerated concrete) with superheated steam at high pressure in an autoclave, converts the materials into Tobermorite, a hydrated calcium silicate. It is only at this stage that the required hardness is achieved.

Figures IV-4.3 and IV-4.4 show SEM images of the green and autoclaved aerated concrete.

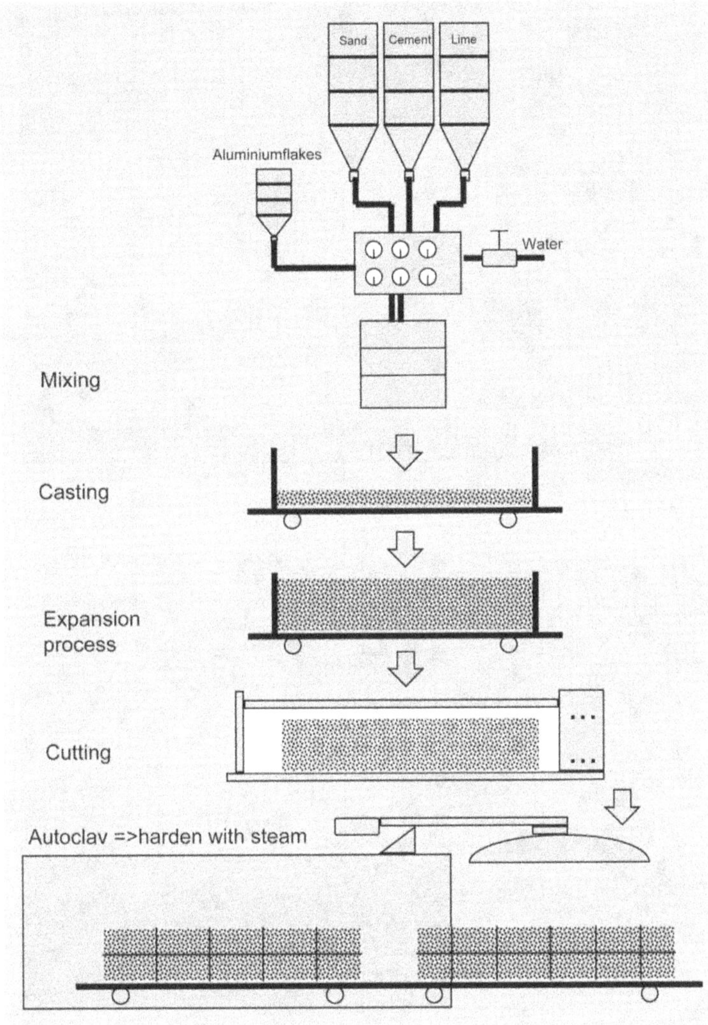

Figure IV-4.2: Production chart of light weight concrete

Technical aluminium flakes

In the search for suitable pore-forming agents, technical aluminium flakes have proved to have a number of advantages. With a density of only 0.0899 g/l hydrogen is a non-toxic, odourless and ecologically safe gas well suited for the blowing process. It is very flammable but is not a major consideration in this industry.

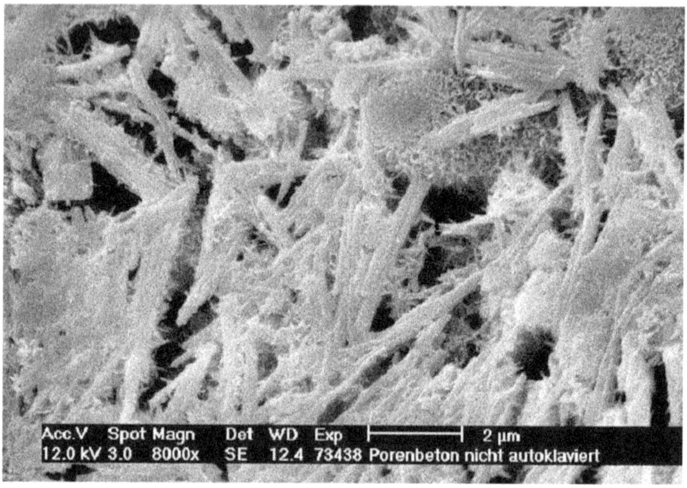

Figure IV-4.3: SEM image of green stage

Figure IV-4.4: SEM image of autoclaved aerated concrete

The variety of technical methods for producing autoclaved aerated concrete and the different strengths with bulk densities (that is density including the pores) from 100 to 800 kg/m^3, make high demands on the metallic products used. The time at which the reaction starts and the variation of hydrogen formation with time must be exactly right.

The technical aluminium grades are used as powders or as pastes. Aluminium pastes are available with water or organic solvents as paste forming agents. They are mostly hydrophilic and this is useful as it allows processing with short mixing times. In practice, addition of special surfactants or heat pre-treatment of the aluminium flakes further improves processability. The reaction properties of the technical aluminium flakes can be adjusted by varying the particle size distribution, the particle coating and inhibition. In general, coarser products are used in the manufacture of higher-density autoclaved aerated concrete; fine particles have proved suitable for grades where lower strength is required.

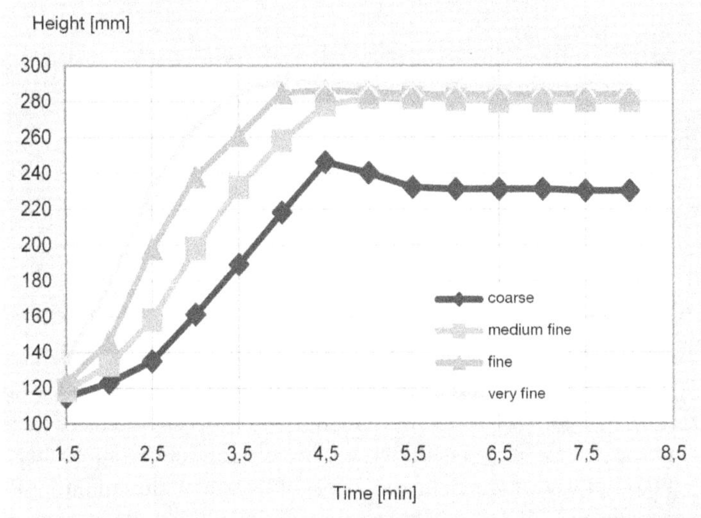

Figure IV-4.5: Reaction performance of different aluminium powders. Laboratory casting of autoclaved aerated concrete-aluminium powder

In Figure IV-4.5 the reaction behaviour of autoclaved aerated concrete powders is shown as a function of the particle size distribution. This uses a special laboratory test to show the expansion during manufacture. Reaction using fine aluminium powders is significantly than with coarser grades. After reaching the maximum size, a decrease in volume is observed with the coarser grades.

Delayed start of reaction

Since the middle of the 1990s, there has been a trend in the autoclaved aerated concrete industry to lower the density for the individual strength grades. For this the expansion process needs to be more accurate and demands a new generation of quality aluminium flakes with different reaction behaviour, particularly in the initial phase of reaction. This has led to the development of so-called technical aluminium grades. This new type of aluminium flake is characterised by, among

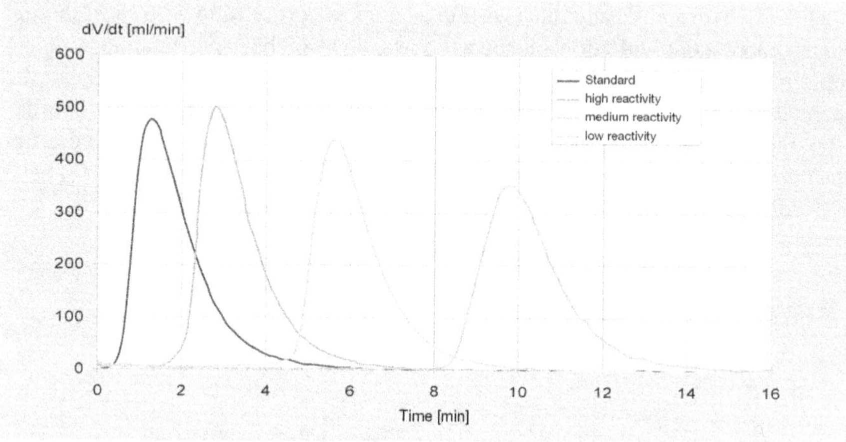

Figure IV-4.6: Hydrogen evolution in saturated calcium hydroxide solution: $Ca(OH)_2$-reactivity of autoclaved aerated concrete paste

other things, the ability to delay the start of reaction for a certain period, which can be as long as several minutes.

Some of the advantages arising from these special grades are the avoidance or at least reduction of hydrogen loss during the mixing process, viscosity increase in the initial phase and the associated increase in void formation (air occlusions). A suitable method of testing the delaying effect of technical aluminium products is the calcium hydroxide reactivity [5] test. In this, the aluminium pigment is reacted with calcium hydroxide solution and the liberated hydrogen measured as a function of time.

Figure IV-4.6 shows aluminium pastes producing a stepwise delay, based on calcium hydroxide reactivity values. Such pastes are in current use and offer autoclaved aerated concrete manufacturers good provision for process optimisation. Table IV-4.3 gives an overview of aluminium flakes used in the autoclaved aerated concrete sector.

Metal flakes in plasters and mortars

External plaster protects masonry from environmental effects such as rain and improves heat and noise insulation. For exterior walls, cement or systems based on lime mortar are normally used as a base rendering. These binders lead to shrinkage that is volume contraction, on setting. Generally the surfaces to be rendered are several square metres in area and large tensile forces are created by shrinkage during setting. These can lead to large cracks in the rendering layer. Such cracks are undesirable from both optical and engineering viewpoints. The imper-

Product	Solvent	Reaction start	Average particle size (d_{50})	Bulk density required
Aluminium powder e.g. STANDART® Aluminium powder lightweight concrete	–	retarded	fine: 24 to 26 µm medium: 43 to 47 µm coarse: 65 to 75 µm	low medium high
Aluminium paste e.g. STAPA® Hydropor	water	standard or retarded	fine: 16 to 20 µm medium: 40 to 45 µm coarse: 70 to 80 µm	low medium high
Aluminium paste e.g. STAPA® Alupor	diethylene glycol	standard or retarded	fine: 16 to 20 µm medium: 40 to 45 µm coarse: 70 to 80 µm	low medium high

Table IV-4.3: Overview of aluminium flakes used in lightweight concrete

fections cannot generally be compensated by the facing rendering which is applied as a final layer. To avoid cracks in the base layer, either the base rendering can be applied in two thinner layers or additives, which compensate for the shrinking, must be incorporated in the render mixture. Applying the rendering in two layers is expensive as twice the amount of work is needed.

Aluminium flakes as an additive

Aluminium flakes have proved suitable as an additive to compensate for shrinkage. Rendering mortars comprise binders, fillers and various additives, for example, water repellents. Now that mortars are no longer made up from the raw materials in situ but are produced commercially as a dry mixture, small amounts of additives must be homogeneously distributed in the mixture. As a dry mixture is to be produced, flake shaped metals must be added in powder form. These mixtures are stored loose in silos or are delivered to the building site in sacks and mixed with water immediately before use. The dry mixtures are formulated so that the gas-producing reaction starts a certain time after the mixture has been applied to the wall and therefore compensates for the shrinkage. The amount of aluminium is calculated so that the shrinkage is compensated for but no significant, undesired volume increase occurs.

Other interesting fields of application for aluminium flakes are renderings where low density or high pore content is required, for example insulating and renovation renderings.

Optical effects

Furthermore, use is made of the metallic gloss of special metallic pigments in high-quality renderings. Following attack of moisture, non-inhibited pigments grey and the desired optical effects are lost, so that such metal pigments must be specially inhibited to prevent those very reactions which in other cases are desired.

Aluminium flakes as expansion agents

A further use of aluminium flakes is in mortars used to fill voids, where not only must the shrinkage be compensated but a volume increase must be obtained. Voids often produce structural problems leading to noise resonance or undesirable accumulation of moisture. To prevent such problems, the voids are filled. So that no new voids form on setting, aluminium flakes are used as an expansion agent. These mortars are also supplied as a dry mixture and mixed with water directly before use.

Compounds

The risk of a dust explosion during manufacture of the dry mortars is avoided by pre-mixing the aluminium flakes with an inert material such as powdered limestone to give commercially available compound. These compounds also allow higher measuring accuracy because larger amounts have to be weighed out.

Reaction rates

The rate of reaction of aluminium flakes in dry mixes depends mainly on the binders and additives, e.g., water repellents, used. Therefore aluminium flakes with

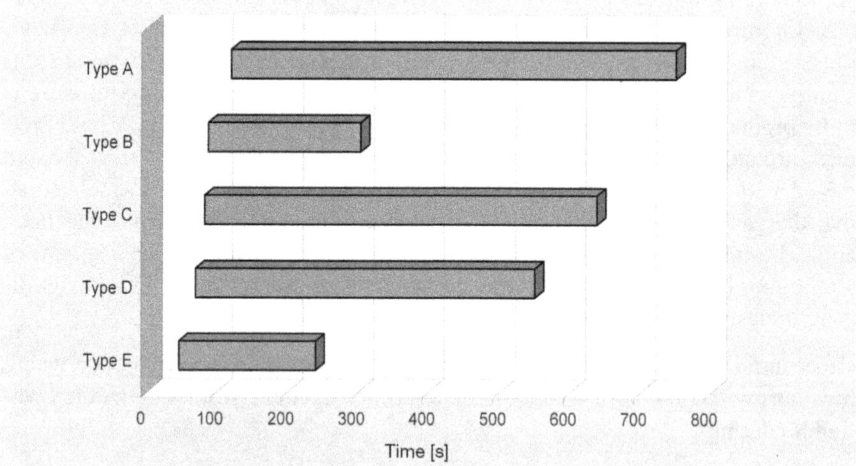

Figure IV-4.7: Reactivity of different aluminium flakes
Reactivity in saturated $Ca(OH)_2$- solution at 25 °C
Start of bar: 10% of the total amount of hydrogen
End of bar: 90% of the total amount of hydrogen

different reactivities are available (Figure IV-4.7). In addition, it is possible to produce products with specially tailored reactivity.

Effect of temperature

In contrast to the manufacture of lightweight concrete, where the temperature can be controlled, with renderings everything depends on the environmental conditions. As shown in Figure IV-4.8, the temperature has a marked effect on gas formation. The reaction rate of the aluminium flakes rises with increasing temperature. However the setting speed of the binder also increases (and with it the stiffening of the mortar) so that, in practice, the effect is not as critical as is suggested by Figure IV-4.8. However an eye must be kept on this parameter as external renderings are generally used in a temperature range from 10 to 40 °C.

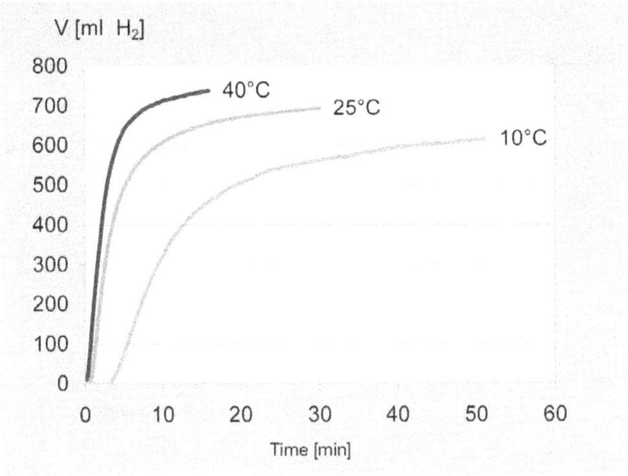

Figure IV-4.8: Liberation of hydrogen at various temperatures (reaction in saturated $Ca(OH)_2$ solution)

In using mortars with metallic powders or pastes as additives to compensate for shrinkage or as expansion agents, care must be taken to ensure that the hydrogen can escape freely so that there is no danger of a hydrogen-air explosion. In exterior use this is normally the case so that no danger exists there.

Metal flakes as sources of energy

Fireworks have a long tradition. Many generations have been thrilled by giant firework displays and the wealth of ideas for new effects seems almost unstoppable. Although we may not be aware of it, we are, from childhood on, conversant with the pyrogenic properties of aluminium whether with sparklers at New Year

Figure IV-4.9: Fireworks with aluminium

celebrations, at open-air festivals or in professional life. Table IV-4.4 indicates some of the most important uses.

Explosives	Explosives Slurries
Pyrotechnic products	Rockets Firecrackers Sparklers Distress rockets Signal lights

Table IV-4.4: Applications of pyro aluminium pigments Pyro/slurry aluminium powder and aluminium glitter

Pyro-products and slurries

In contrast to technical aluminium flakes as pore-forming agents, which are generally wettable by water, pyro-products contain a certain amount of fat. The fat content of slurry products is particularly high, sometimes over 4%. Slurry explosives are among the safety explosives. They are moist or liquid or gel-like, dust-free, easy to handle and to a certain extent pumpable. Addition of technical flakes, also known as slurry aluminium flakes, mainly serves to sensitise the explosive. To retain the reaction properties of the aluminium the particles are often stabilised with saturated fatty acids. The demands made on pyro-aluminium flakes are many and depend strongly on the area of application.

Particle size distribution

Pyro powders with a particle size less than 45 µm (> 90 w/w) are considered fine products. "Coarse" refers to grades which have an average particle size over 100 µm. Moreover, the colour can be a criterion with the available palette ranging from silver grey to very dark, almost black. In special cases low-dust, pyro-powders are used. The low-dust property is obtained by appropriate treatment of the pigment surface.

Testing the reactivity

The so-called burning test [6] permits reactivity to be assessed. Pyro-aluminium powder is mixed with an oxidant, usually potassium perchlorate, and the mixture burnt in a cardboard tube. The duration of burning is measured – with very reactive powders this is under 7 seconds (Figure IV-4.10). In some applications only a small volume of the pyrotechnic mixture is available. In such cases pyro-aluminium with relatively high bulk density of over 0.50 kg/l is used. Mixing can also be favoured by heavier grades.

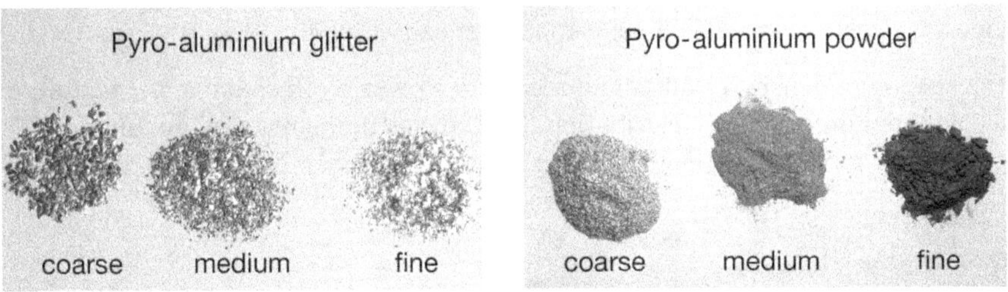

Figure IV-4.10: Pyro aluminium of different fineness

Use as a propellant

Pyro-aluminium powder is often found in propellant mixtures. In this high-speed, but non-explosive, combustion is desired. The requirements for a controlled reaction are satisfied by aluminium grades as they do not completely react in the primary detonation but only later in a secondary stage. This leads to "post heating" and the propellant effect.

Aluminium glitter

It is not only the extremely thinly milled aluminium flakes with high specific surfaces of over 20,000 cm^2/g which are used. Aluminium glitter, that is flakes with average particle size over 150 µm, are also among items needed for display fireworks. Aluminium glitter is available in different particle sizes ranging from 150 µm to ca 4,000 µm. The manufacture of narrow particle size distributions is technologically difficult because of the surface structure of the particles. For completeness it should be noted that aluminium is also used in granular form in pyrotechnics.

Metal flakes in the chemical industry

Aluminium products, particularly aluminium flakes are used in various processes in chemical industry. As already mentioned it is the high reduction potential of the reactive metal which is important. Reactions always occur at a phase boundary, that is at a solid/liquid, solid/gas or solid/solid interface. It is necessary to produce the metal with appropriately large surface area. A small particle size permits more uniform distribution in the reaction medium.

Aluminium flakes as a reducing agent

An important chemical process in which the flakes are used as a reducing agent is the manufacture of titanium dioxide pigments by the sulphate process. In this process, aluminium flakes are used to reduce iron from tri- to di-valent state. Only after this reduction can the iron be separated, a step which is necessary to obtain the pure white colour. In these processes, aluminium flakes have definite advantages over zinc flakes which are also theoretically usable:

- The lower density of aluminium leads to slower sedimentation and therefore a more homogeneous distribution of the flakes in the reaction mixture.
- Weight-for-weight about 3.5 times as much hydrogen is produced.

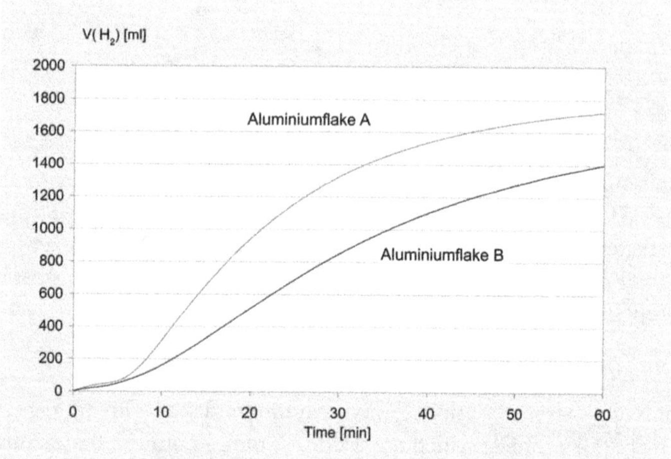

Figure IV-4.11: Hydrogen production by two typical aluminium flakes at 50 °C in 10% sulphuric acid. Hydrogen liberation is still not complete after 60 minutes.

- The reaction product, aluminium oxide, is often used in small amounts in the manufacture of titanium dioxide to improve the properties of the pigment so that the reaction product does not impair the subsequent processes.

For chemical uses, aluminium flakes are available with various particle sizes and surface modifications and these allow the start and duration of reactions to be varied by a certain amount. Material consumption and yield can be optimised. Figure IV-4.11 shows the course of hydrogen formation in the manufacture of titanium hydroxide by the sulphate process for two typical products in 10% sulphuric acid. For all processes in solution, aluminium flakes formed into a paste with organic solvent or water are recommended. This makes dispersion easier and the risk of explosion is avoided when processing is carried out correctly.

References

[1] Eckart test procedure: Document 6C.0421.0
[2] Eckart test procedure: Document 3C.4002.0
[3] Eckart test procedure: Document 6C.3411.0
[4] *Stieß, Matthias:* Mechanische Verfahrenstechnik 1. Berlin/Heidelberg 1992; p. 123ff
[5] Eckart test procedure: Document 6C.4005.0
[6] Eckart test procedure: Document 6C.3501.0

5 Corrosion protection

Corrosion, the chemical or electrochemical reaction of a metal with its environment, is causing huge economic costs and consequential damage. It is not without reason that in the automotive industry, for example, 10 to 20% of the total expenditure goes on preventive anti-corrosion measures. Corrosion protection maintains value and plays an important role in safety, etc.

The anti-corrosive effect of zinc dust, that is isometric zinc particles, has been known for decades. In general the most common anti-corrosive coating is zinc rich primer produced according to DIN 50976 (SAE-AMS-P-38336) with at least 90% zinc in the dry film. Thousands of tons of zinc dust are used worldwide each year to protect capital goods, steel structures and household articles (cf. p. 142 f).

Flake-shaped zinc pigments offer an alternative to zinc dust and can produce comparable corrosion protection with significantly less zinc in the dry film and consequent cost savings.

Mode of action of flake-shaped zinc pigments

Development of zinc pigment flakes started in the 1970s initiated by the desire for lower film thickness in coatings which maintain the excellent corrosion protection when subject to mechanical loading. For manufacturers of lamellar metallic pigments it was obvious to develop zinc pigment flakes. Zinc flakes are manufactured by wet grinding with a lubricant in ball mills similar to aluminium pigments or by dry-grinding as for gold-bronze pigments (cf. Part II, Chapter 2).

Zinc occurs in nature mainly as the sulphide or oxide. Zinc blende, a zinc sulphide, is the most common ore used for smelting. It is found in Poland, France, England, Australia, Canada, and the USA. It is first reduced and subsequently purified electrolytically.

In addition to zinc metal providing cathodic protection against corrosion, zinc flakes also have the advantage of excellent barrier protection. The barrier effect provides long-term protection while cathodic protection against corrosion is only temporarily. The mechanism of protection involves so-called cementation in which, as a result of zinc oxide formation, particles grow in size sealing up any pores and covering damaged areas.

Cathodic protection against corrosion requires a conductive contact both between zinc and the steel surface and between the zinc particles themselves. With zinc dust, point contacts with relatively high resistance are made between the isometric particles but zinc flakes have a more extensive contact which allows better electrical conductivity and consequently better anti-corrosive properties – although this also leads to faster degradation of the zinc flakes. In Table IV-5.1 technical data for zinc flakes and zinc dust are compared.

Attribute	Zinc dust	Zinc flake
Particle shape	isometric	platelet, flake
Specific weight	approx. 2.2 kg/l	approx. 1.1 kg/l
Oil absorption	approx. 6.5 g/100 g	approx. 22 g/100 g
Average particle size	approx. 3 µm	approx. 15 µm
Specific surface	approx. 0.5 m^2/g	approx. 3 m^2/g
Absorption of binder	low	high
Pigment volume concentration	high	low

Table IV-5.1 Comparison of technical data on zinc flakes and zinc dust

To counteract the quicker breakdown of the zinc flakes, flake-shaped fillers (e.g. mica) and/or anti-corrosive pigments such as zinc phosphate are introduced into the formulation. By combination with aluminium pigment flakes, the reactivity of the zinc flakes is lowered and the period of cathodic protection extended. A typical mixing ratio is one part of aluminium pigment to three parts of zinc flake (solid pigment by weight). Addition of mica to zinc dust mainly serves to reduce the tendency for sedimentation. As the electrical conductivity must be preserved, the amount of filler which can be used is small. In addition to extending the period of anti-corrosive activity, mica pigment combined with zinc flakes also lowers the raw material costs.

Processing, application and coating properties

Because of their flake structure, zinc flakes have a significantly higher specific surface area than isometric zinc dust. The correspondingly higher binder uptake is of prime importance for formulations with lower pigment volume concentration (PVC). Consequently this provides advantages for both application (brushing, pneumatic and airless spraying, dipping and rolling) and paint properties.

The main advantages are:

- higher formulation flexibility
- better substrate- and inter-coat adhesion
- good overcoatability
- low porosity and permeability
- smooth surface
- metallic appearance
- significantly less tendency to settle
- loose, easily re-stirrable sediment

	Salt spray cabinet DIN 50021 / ISO 9227		Condensation test DIN 50017/ ISO 6270		Alternating Climate Test 00 according *Vorschrift Deutsche Bahn* KSW 2.0 S		Kesternich Test DIN 50018 according *Vorschrift Deutsche Bahn* KSW 2.0 S		Haagen Test	
	687.03	ZnFl 9/58	687.03	ZnFl 9/58	687.03	ZnFl 9/58	687.03	ZnFl 9/58	687.03	ZnFl 9/58
degree of blistering DIN 53209 / ISO 4628	m0 g0	m0 g0	m0 g0	m0 g0	m0 g0	m0 g0			m0 g0	m0 g0
visual inspection	corrosion at the cut no undercreep	little corrosion at the cut no undercreep	no corrosion	no corrosion			no corrosion	no corrosion	no corrosion	no corrosion
cross cut DIN 53151 / ISO 409	3 Gt 1	3 Gt 0	3 Gt 0-1	3 Gt 0	3 Gt 0-1	3 Gt 1	3 Gt 0	3 Gt 0	3 Gt 1-2	3 Gt 1-2
undercreep [mm]					Wb 0	Wb 0				
degree of corrosion DIN 53210 / ISO 4628					Ri 0	Ri 0				

Table IV-5.2: Results of corrosion protection tests (formulations 687.03 and ZnFl 9/58 are explained in the text)

Results of anti-corrosion tests

Table IV-5.2 shows results of individual tests on the anti-corrosive effect of formulation ZnFl 9/58 based on zinc flakes (PVC 60% and film thickness 70 μm) compared with a standard formulation based on zinc dust (PVC 90% and film thickness 70 μm according Sheet 87 "Technische Lieferbedingungen TL 918 200 Teil 2" of Deutsche Bahn (German Railways), material number 687.03).

The test results show that the anti-corrosive effect of both formulations is practically identical. Some properties of the anti-corrosive paint with lamellar zinc are better. Anti-corrosive paints based on zinc flakes have also proved their value in outdoor weathering tests (sea and mountain climates) and in practical application (railway bridges) [1].

Areas of application

Protection against corrosion with zinc based systems is both sensible and necessary in many areas. Of course, flake shaped zinc pigments cannot replace tried-and-tested zinc dust or galvanisation in all applications. This is particularly true for large areas which are to be coated economically.

For years the use of zinc flakes has proved itself mostly in inorganic coatings for mass-produced items. The items to be coated vary greatly from screws, nuts and springs, through tubes to complete exhaust systems. Depending on the shape and size, application of anti-corrosive coating can be done by classical dipping (e.g. tubes), in the dip/spin process (e.g. small goods in drums) or by electrostatic spraying.

Use of zinc flakes in heavy-duty corrosion protection is possible without problems but acceptance in practice is very slow. In many cases the introduction of zinc flakes is blocked by existing manuals and standards which are based on (very high) minimum zinc amounts, that can only be achieved with zinc dust. The most commonly used standard in Germany, and possibly in Europe, is the specification "Technical Delivery Conditions of Deutsche Bahn". Tests in practice of a zinc-flake-based coating have already been completed successfully [1]. The first commercial applications for zinc flakes are in shop- and container-primers. In these applications the degree of corrosion protection to be achieved is specified, the composition of the dry film is not part of the specification.

Environment and economics

Currently there is an ongoing discussion where zinc is regarded as a heavy metal with some ecologically damaging effects. It is therefore necessary to respect also ecological aspects (next to technical considerations) which means reducing the amount of zinc in use. Table IV-5.3 shows the amounts of zinc used in a zinc

Zinc consumptioon and cost	Formulation with zinc dust (as reference) according "Material-Nr. 678.03"	Formulation with zinc flake according "Material-Nr. ZnFl 9/58"
g zinc/kg in the anti-corrosion coating	approx. 750 g/kg	approx. 197 g/kg
Coated area: m^2/kg anti-corrosion paint with 70 µm film thickness	2.63 m^2/kg	4.17 m^2/kg
Paint consumption, g zinc/m^2 anti-corrosion coating	approx. 286 g/m^2	approx. 47 g/m^2
Material cost per m^2 (70 µm)	basic price	as 687.03
Cost per m^2 (70 µm) for handling (transport, weight, application etc.)	reference price	ca. 20% below reference due to better handling and other cost saving options

Table IV-5.3: Comparison of amounts of zinc used (Formulations 687.03 and ZnFl 9/58 as in Table IV-5.1)

based standard formulation made up as specified in Sheet 87 (Blatt 87) compared to an alternative formulation with zinc-flake.

The lower zinc content per kg of primer and the lower zinc amount required per m^2 of covered surface when using zinc pigment flakes compared with zinc dust primers offers significant ecological advantages. Further to that, there is also a positive effect on the cost – taking into account not only the costs of the materials but also the expenditure on handling (lower transport costs because of lower weight, and better and quicker stirrability and application), the zinc-flake based formulation is also more favourable economically.

For the reasons given (cf. p. 101 f), use of zinc flakes in water-based systems is at present of secondary importance.

Concluding remarks

To summarise, it is fairly obvious that the anti-corrosive properties of standard formulations based on zinc dust and formulations based on zinc flakes are practically identical and that there are technical and ecological advantages when using zinc flakes without any economic disadvantages. For coatings manufacturers and users there are particular advantages of zinc flakes such as better handling, more freedom in formulation, easier application and lower material consumption per covered area.

References

[1] *Fetz, A.; Gentz, E.; Gessner, H.-J.:* Korrosionsschutz mit Zinkflakes. Blättchenförmige Korrosionsschutzpigmente bieten eine umweltgerechte und wirtschaftliche Alternative. Farbe & Lack, November 1997, pp. 118-131

Guiding formulation
Tripigment zinc primer based on 2C-epoxy resin
air drying Rec. No. 22-07003

ECKART

Araldite GZ 7071 X 75 BD	13.50	www.huntsman.com
Solvesso 100	21.45	www.exxon.com
MIBK	7.80	
Butanol	3.70	
Anti-Terra 203	0.80	www.byk-chemie.com
Thixatrol ST	0.40	www.elementis-specialties.com
Blanc fixe micro	12.30	www.sachleben.de
Micro Talkum AT extra	15.90	www.omya.com
Heucophos ZCP 3,00 disperse with dissolver	3.00	www.heubachcolor.de
Standart Zink AT mix with dissolver at low shear rate	21.15	www.eckart.net
	100.00	

Hardener:	Aradur 115 X 70 BD (www.huntsman.com)
Mixing rate:	100 p.o.w. of primer 7 p.o.w. of hardener
Diluent:	Solvesso 100 / Butanol 4:1
Spraying viscosity:	app. 20 sec. in DIN cup 4 mm
PVC:	app. 63 %

6 Cosmetics and personal care products

Effect pigments give beauty and elegance to all types of products, especially in the area of colour cosmetics and personal care. The unique beauty of metallic effect pigments is based on strong reflection of the incident light beams from the surface of flakes. Following the strict restrictions of the cosmetics regulations, only a small selection of metallic effect pigments can be used in cosmetics. Aluminium, bronze and copper are allowed worldwide and these pigments allow corresponding silver and gold to copper colours to be produced. Depending on the particle size, pigment loading and formulation a variety of effects can be produced from subtle sheen, brilliant gloss or sparkling reflection.

The metallic effect and the silver, gold and copper colours can be successfully combined with other colourants, to create new colours and effects not otherwise possible in cosmetics. In combination with pearlescent pigments and/or organic pigments unique polychromatic effects can be produced. The metallic effect is greatly impacted by the pigment orientation in relation to the surface and smoothness of the substrate, depending considerably on the formulation, viscosity and application conditions. Metallic effect pigments are by nature opaque. The excellent hiding power or coverage compared with other effect pigments, for example pearlescent pigments, has a number of advantages in colour cosmetics for both the formulator and user. In formulating good coverage in lipsticks, eye shadow or nail varnish can be achieved with low metallic pigment loading.

Fields of application

Uncoated aluminium-, bronze- and copper pigments can be used comprehensively in anhydrous cosmetics and personal care formulations. In contrast, in water-based systems the use of metallic effect pigments is very limited. Thus aluminium pigments can react with water forming hydrogen which leads to greying of the pigment. Bronze- and copper pigments alter their colour in water-based systems by oxidation of the copper.

Thanks to a new stabilised generation of metallic effect pigments it is now possible to extend the use of metallic effect pigments to water-based systems. Passivation is obtained by encapsulation of the metallic effect pigment in silica (cf. p. 88 f). Eckart manufactures this type of pigment especially for the cosmetic and personal care area. These metallic pigments range contains 14 products in nine different colours and in two different particle sizes (Table IV-6.1). They are specified in the cosmetics industry using INCI nomenclature (International Nomenclature for Cosmetic Ingredients) as aluminium powder, bronze powder, copper powder and silica. The high purity of the metal and the vegetable-based milling aids satisfy all requirements of the cosmetics regulations.

Product	Average particle size (d_{50})	Delivery form	Opacity	Brilliance	INCI name
Aluminium pigments	17 or 35 μm	powder	good to very good	good to very good	
e.g. VISIONAIRE® Bright Silver Sea	35 μm	powder	good	very good	aluminium powder and silica
Bronze pigments	17 or 35 μm	powder	good to very good	good to very good	
e.g. VISIONAIRE® Bright Natural Gold	35 μm	powder	good	very good	bronze powder and silica
e.g. VISIONAIRE® Sunflower Gold	17 μm	powder	very good	good	bronze powder and silica
e.g. VISIONAIRE® Amber	17 μm	powder	very good	good	bronze powder and silica
Copper pigments	17 or 35 μm	powder	good to very good	good to very good	
e.g. VISIONAIRE® Bright Cinnamon	35 μm	powder	good	very good	copper powder and silica
e.g. VISIONAIRE® Lava	17 μm	powder	very good	good	copper powder and silica

Table IV-6.1: Metallic effect pigments for cosmetics

Formulations

The stability of the metallic pigments in individual applications can depend strongly on the other ingredients, processing conditions, and order of addition within the formulation. The recommended pH is between 6 and 8. Raising the pH should preferably be done with very diluted amines such as aminomethylpropanol, or diluted caustic soda. Lower pH can be achieved with very diluted acetic or citric acid. With emulsions and other water-containing systems, the best results are obtained if the pigment is first added to the oil phase followed by homogenisation using minimal shear and a minimal dwell time. In some applications, the orientation and metallic effect can be improved by adding lecithin as a wetting agent. Depending on the use and the desired effect, the pigment loading should be between 2 and 15%.

Lipsticks, cosmetic pencils, hot-pour foundations, hot-pour blushes, and lip pots

A loading level of 5 to 15% metallic pigment will give some unique and attractive polychromatic effects when mixed with organic pigments or with a combination of organic and pearlescent pigments. Even low loading levels of metallic pigments can improve coverage of the lipstick, foundation or blush. With normal processing and cooling rates, settling of the pigments typically is not a problem. However, if settling should become of concern, a small quantity of trihydroxystearin (e.g. 0.25 to 1%) or tribehenin can provide suspension of the pigments.

Aluminium is currently not permitted for lip products in USA. Aluminium is permitted for lip products in the European Union as well as Japan.

Pressed powders: eye shadow, blush, foundations

Any of the metallic pigments can be used for beautiful polychromatic effects. It is important to not over-shear the metal platelets. Therefore metallic pigments should be mixed at as the last step (along with pearlescent pigments if used). The metallic pigments are much easier to press compared to pearlescent pigments. Again the outstanding coverage with metallic pigments can be a distinct advantage in these formulations.

Mascara

For emulsion based mascaras, it is recommended to use the silica encapsulated metallic pigments, preferably adding it to the oil phase. For anhydrous mascaras, either the standard cosmetic grade pigments or the encapsulated Visionaire® metallic pigments will work.

Emulsions: foundations, creams and lotions, skin care products

For emulsion based formulations it is necessary to use the Visionaire® encapsulated pigments. The silica encapsulated metallic pigments should be added to the oil phase for best results. The homogenization stage should be kept at a minimum for best results. The colour stability from the metallic pigments can be formulation dependent, especially for the bronze and copper shades, and stability tests should be performed during evaluations.

Clear cosmetic gels

For best results, add the silica encapsulated metallic pigments to the alcohol or glycol, if present, before adding in the water. If metallic pigments are desired in body gels, or other clear water-based formulations, suspension of the pigment becomes important. There are some clear water-based thickeners with good yield value and therefore good suspension properties. It is important to know that the thickener of choice will depend on the pH, salt content, presence and amount of water-miscible solvents as well as other ingredients in the formulation.

Perfumes and fragrances

Anhydrous perfumes tend to be stable with aluminium, bronze and copper. For colognes or other fragrances that contain water, the silica encapsulated metallic pigments will need to be used.

Nail polish

Aluminium pigments are widely used in nail polish. It can add some sparkle and flop effects when using the larger particle sizes. The addition of aluminium results in polychromatic effects when combined with organic and/or pearlescent pigments. Traditionally bronze and copper have not been able to be used in nail polish, since nitrocellulose will crosslink in the presence of these pigments. It is believed that the copper ion is involved with this reaction. Depending on the acid value of some base formulations, the Visionaire® encapsulated bronze and copper can provide sufficient stability to be used in nail lacquers. A low acid value resin base has the best stability in nail lacquers.

METALURE® pigments are microscopically thin aluminium platelets manufactured in a physical vapour deposition process (PVD). This guarantees pigments with uniform thickness and a homogenous smooth surface. PVD aluminium dispersions offer extraordinary brilliance and hiding power, far beyond that which can be achieved with conventional aluminium pigments. PVD aluminium dispersions supplied in ethyl acetate are suitable for all nail polish systems and are completely free of any lubricants or other additives. Nail lacquers based on PVD aluminium dispersions are ideally suited for tinting to achieve mirror-like polychromatic effects

Product	Average particle size (d_{50})	Delivery form	Opacity	Brilliance	INCI name
PVD-aluminium pigment dispersion	10 to 13 µm	dispersion	excellent	excellent	
e.g. METALURE® C-1100	13 µm	dispersion	excellent	excellent	aluminium powder and ethyl acetate

Table IV-6.2: PVD aluminium dispersions for cosmetics

The brilliance of PVD aluminium dispersions can be affected by certain solvents and co-solvents and very much depends on creating the right formulation. Suspension of pigments in nail polish is achieved with stearalkonium bentonite or stearalkonium hectorite at 0.2 to 2% loading levels. These suspension agents do require high shear dispersion to properly incorporate to build the physical network.

Hair care

Metallic pigments offer some possibilities in hair care, particularly for the youth and teen market. Hair mascara, hair sprays and temporary hair dyes have used bright colours and special effects for this market segment. Colouring mousses or colouring gels are other possibilities for the Visionaire® pigments for styling products that impart colour highlights. As with other emulsions, it is recommended to add the Visionaire® to the oil phase for temporary hair colour and hair mascara. It is recommended to add the Visionaire® to the alcohol or solvents in hair sprays, colour mousses or other styling products for best results.

Product category	USA	EU	Japan
External/skin care	approved	approved	approved
Eye cosmetics	approved	approved	approved
Lipstick/oral	not approved	approved	approved
Demake up products	approved	approved	approved

Table IV-6.3: Approved applications of aluminium pigments

Product category	USA	EU	Japan
External/skin care	approved	approved	approved*
Eye cosmetics	approved	approved	approved
Lipstick/oral	approved	approved	approved
Demake up products	approved	approved	approved
* according the statute, valid from April 1st of 2001, Japan orients by the established regulations of FDA			

Table IV-6.4: Approved applications of goldbronze and copper pigments

Product safety

Visionaire® and Metalure® C metallic effect pigments from Eckart have been specially developed to suit the demands of the cosmetics market. They are manufactured using the purest metals and at no time during the process do they contain anything of animal origin. The approval data are summarised in Tables IV-6.3 and IV-6.4. Eckart metallic effect pigments for cosmetics applications conform to the requirements of the FDA 21 US Code of Federal Regulations Part 73 Section 73.1645, 73.1646, 73.1647 for Drugs und Section 73.2645, 73.2646, 73.2647 for Cosmetics. Eckart cosmetic aluminium pigments are manufactured according to the criteria of the EU guideline 95/45/EC concerning purity of food colourants.

Furthermore, Visionaire® metallic effect pigments were successfully tested on the skin of people without evidence of irritation in the Repeat Insult Patch Test (RIPT) and eye irritation was checked using an in-vitro test.

References

[1] *Hostynek, J.; Maibach, H.:* Tin, zinc and selenium: metals in cosmetics and personal care products. Cosmetic & Toiletries, 2002, 117 (9), pp. 32-42

[2] *Hostynek, J.:* Chromium, cobalt, copper and iron: metals in personal care products. In: Cosmetic & Toiletries, 2002, 115 (8), pp. 52-65

[3] *Hostynek, J.:* Metals in personal care products. In: Cosmetic & Toiletries, 1999, 114 (3), pp. 47-56

[4] *Kemper, F. H.; Luepke, N. P.; Umbach, W.:* Blue List Cosmetic Ingredients. Aulendorf 2000

[5] The Merck Index: An encyclopedia of chemicals, drugs and biologicals. Rahway, NJ 198310

[6] *Cassano, N.; Amoruso, A.; Vena, G. A.:* Zinc nutritive and skin: An overview. In: J. Appl. Cosmetol, 2002, 20 (3), pp. 183-94

Guiding formulation ECKART

1. Bronze and pearlescent lipstick

INCI Name	% w/w
Phase 1	
Carnauba wax	3.0
Beeswax	2.5
Candelilla wax	5.0
Ozokerite	2.0
PTFE	2.0
Castor oil	17.82
Polyisobutene	7.0
Isostearyl isostearate	5.0
Isononyl isononanoate	18.0
Caprylic/capric triglycerides	10.0
Phase 2	
FD & C Red 40 Aluminum lake	10 33% dispersion in castor oil
FD & C Blue 1 Aluminum lake	0.03 50% dispersion in castor oil
Phase 3	
Propyl paraben	0.10
BHT	0.05
Prestige® Bright Fire-Red	
Mica (and) iron oxides	7.5
Visionaire® Bright Natural Gold	
Bronze powder and silica	7.5
Boron nitride	2.5

Procedure

Phase 1: mix with low shear, heat to 80 to 84 °C

Phase 2: add and mix until completely homogenised

Phase 3: add and fill into a lipstick case at 70 °C

Mix phase 1 with a low shear mixer
Heat to 80 to 84 °C, under mixing
Add phase 2, mixing until homogenous
Cool to 70 to 75 °C, and add phase 3
Pour into a lipstick mold at 70 °C

Guiding formulation	ECKART
2. Bronze and pearl pressed eye shadow	
INCI Name	**% w/w**
Phase 1	
Mica	34.9
Titanium dioxide	5.5
Chromium oxide green	2.8
Iron oxide yellow	1.8
Zinc stearate	5.5
Boron nitride	1.8
Phase 2	
Prestige® Sparkling Lemon Gold	
Mica (and) titanium dioxide (and) iron oxides	18.35
Visionaire® Bright Natural Gold Bronze powder and silica	18.35
Phase 3	
Dimethicone	5.5
Caprylic/capric triglycerides	5.5

Procedure

Mix phase 1 in a blender until uniform
Check pigment dispersion
Add phase 2 and mix until uniform
Spray in phase 3 and mix until uniform

Guiding formulation ECKART

3. Nail varnish with PVD pigments

INCI Name	% w/w
METALURE® CL 51012 AE	2.00
Nitrocellulose	15.00
Ethyl acetate	19.40
n-Butyl acetate	38.70
Isopropyl alcohol	8.00
Polyester resin	9.00
Acetyl tributyl citrate	7.00
Stearalkonium bentonite	0.90

Procedure

Disperse stearalkonium bentonite in butyl acetate under high shear. Then activate with a mixture of 2:1 mixture of IPA/H_2O at 30% of the weight of the stearalkonium bentonite. Add the rest of the ingredient except Metalure and mix well.
Add the Metalure dispersion under minimal shear and mix until uniform.

Part V Effects on people and environment

Physiological effects

Based on many years of experience, Eckart STANDART® gold bronze, aluminium and zinc pigments are regarded as non-toxic. Inhalation of the vapour of boiling copper can cause "metal vapour fever". Copper-containing dust may cause similar symptoms if it is present in the working environment in very high concentrations, above the upper limit for industrial exposure [1]. Chronic poisoning by gold bronze, aluminium and zinc pigments is not known [2].

Approved areas of application for metallic pigments

Cosmetics

Gold bronze and aluminium pigments of the VISIONAIRE® series are approved for use in cosmetics in the European Union by the Cosmetics Directive 76/768/EC and in the USA by Title 21 CFR § 73 ff.

Food packaging

STANDART® gold bronze and aluminium pigments conform to the requirements of the IXth Recommendation of the German BfR. STANDART® aluminium pigments are also permitted in the European Union according to Plastics Directive 2002/72/EC and in the USA according to Title 21 CFR § 175.300 and § 178.3297.

Toys

STANDART® gold bronze and aluminium pigments satisfy the requirements of the European Toy Standard EN 71 part 3.

References

[1] *Merian, E.:* Metalle in der Umwelt. Weinheim 1994
[2] ETAD; BCMA; VdMI; EPSOM: Safe Handling of Pigments. 1995

Part VI Safety advice

Metallic effect pigment pastes and powders can be safely transported, handled and stored as long as some basic rules and principles are followed. As the potential for danger of the various powder pigments, product groups and preparations differ greatly from each other, correct assessment and handling always requires careful consideration of the safety data sheet of each individual product. For example, preparation of a solvent-containing pigment paste greatly reduces the dangers from dust formation but, under some circumstances, the dangerous properties of the solvent must now be taken into account. For a safe working environment with both pastes and powders, the tidiness of regularly cleaning the relevant production areas is of particular importance.

Physical hazards of metallic effect pigment powders and pastes

The hazards of aluminium and zinc metallic effect pigments are:

- flammability
- formation of explosive dust/air mixtures in case of powder dust clouds
- formation of flammable hydrogen on contact with acids or bases and on long-term contact with water. Together with air (in certain proportions), hydrogen forms an explosive mixture.

The hazards of gold-bronze pigments (copper or copper/zinc alloy) are:

- flammability
- pure copper is classified as marine pollutant (transport regulations for dangerous goods)

These physical properties are strongly affected by the specific surface area of the particles and therefore by the fineness of the product. For example, metal blocks (ingots) of aluminium or zinc are not representing any hazardous potential unless they are exposed to acids or bases or any high temperatures above the melting point (e.g. in case of fire). However, fine aluminium powder has a much higher tendency to react because of its much greater surface area: this needs to be considered in safety precautions.

Most important, very fine aluminium powder can be easily whirled up. If the air metal dust cloud comes into contact with an ignition source, it results in very rapid combustion in form of explosion.

Stirring up aluminium or zinc dust should be avoided strictly. Safety regulations against explosions, require potential sources of ignition (sparks, hot surfaces, electrostatic discharges) to be avoided and nitrogen (inert atmosphere) to be applied where appropriate.

As with all fine-particled products, aluminium and zinc pigment powders can become electrostatically charged by friction during handling in production. This can lead to sparks. To exclude this hazard, the relevant processing equipment must be earthed. Earthing of containers and equipment, for example using earthing clamps, when filling or emptying, is also essential for certain products. When transferring and filling aluminium or zinc powder, only the original metal container must be used: no plastic containers!

By using solvents as pasting agents, the formation of dust and the risk of dust explosion is greatly reduced or completely eliminated. However the type of solvent used can give rise to new hazards, especially in case of irritating or flammable solvents.

For this reason, the safety data sheet for the particular product or paste preparation should be closely studied as the information contained is so comprehensive that, if followed, both the user and the environment will be safely protected. Alongside a description of the product and data on composition, potential risks, first-aid measures, handling and storage, control of exposure and personal protection, physical and chemical properties, stability and tendency to react and legal requirements, all the other information necessary for safe use of the product is given.

Fire fighting when metallic effect pigment powders and pastes are involved

Fire involving metal pastes normally starts as a solvent fire and in the initial stages can be fought with normal ABC-Extinguishing powder. If this is unsuccessful or if the fire intensifies this means that the fire has become a metal fire. It can now only be successfully fought with special metal fire extinguishing powder (fire class D) or with dry sand.

Metal powder fires must immediately be tackled with metal fire extinguishing powder or best of all with dry sand. There should always be a sufficient quantity of dry sand and shovels available to immediately form a boundary around the source of fire and cover it. Covering with sand must be done very cautiously using long-handled shovels. Stirring up the burning powder must be strictly avoided, as a serious explosion is possible with only a small dust cloud.

All liquid, halogenated, carbon dioxide or foam extinguishing agents must be strictly kept away from areas used for storage or processing of dry powders so as to avoid their use by excited and/or inexperienced people during an emergency.

Water must never be used to try to extinguish a burning metal powder as the hydrogen formed will fan and spread the fire. Information and documentation on the correct way to tackle metal fires should also be made available to external emergency services.

Special points concerning handling of aluminium and zinc pigment powders

When handling aluminium and zinc powders, electrostatic spark discharges, particularly with fine-particled pigments, can cause ignition of the metal dust/air mixture. Care must therefore be taken to ensure that all equipment and utilities such as mixers, shovels, containers, etc., are earthed. Furthermore, no tools made of iron may be used as the rust which forms can cause sparks. Personnel working in the factory should be equipped with flame retardant, conducting clothing, safety shoes, gloves and a helmet with a safety visor. Only the amount of aluminium necessary for the day should be stocked in the working area. The pigments should be filled into conductive metal containers not into plastic drums, even if these have been certified as conductive by their manufacturer. It must be ensured that no open flames are in the vicinity. Furthermore, hot surfaces, such as occasionally occurring on damaged shafts and bearings, are to be avoided.

General safety measures if working with metallic effect pigment powders and pastes

In addition to using the recommended personal safety equipment, it is important to lay down suitable rules for working and personal hygiene. Eating, drinking, smoking and the use of cosmetics are not allowed in the work area. Before any of these activities, people must wash their hands. Machines delivering food, drink or cigarettes may not be set up in the work area. Work clothes must be kept separate from normal clothes. Storage and handling of dangerous materials can only be performed by special trained personnel, exclusively taking place in clearly marked areas. The workplace must always be kept clean and tidy. Regular instruction of personnel about possible risks in dealing with dangerous materials is particularly important. Furthermore, staff must always be encouraged to follow good working practice

References

[1] Ullmanns Enzyklopädie der technischen Chemie: Anorganische Pigmente. Band A20, 1992

[2] *Merian, E.:* Metalle in der Umwelt. Weinheim 1994

[3] Technische Regeln für Gefahrstoffe TRGS220. Ausschuss für Gefahrstoffe, April 2002

[4] National Fire Protection Association (USA): Prevention of fire and dust explosions in the chemical dye, pharmaceutical and plastics industries. NFPA 654, 1988

[5] *Bartknecht, W.:* Dust explosions – cause, prevention, protection. Berlin/Heidelberg/New York 1989

[6] National Fire Protection Association (USA): Machining and finishing of aluminium and the production and handling of aluminium powders. NFPA 651, 1998

[7] The Aluminium Association: Recommendations for storage and handling of aluminium powders and paste. Washington D.C. 20004

Authors

Kapitel		Autor
	Foreword	Dr. Klaus Greiwe
I	Historical background	Peter Wißling
II	Fundamentals	
II-1	Metallic effect pigments	Peter Wißling, Dr. Alfried Kiehl
II-2	Manufacture and properties of aluminium, goldbronze and zink pigments	Peter Wißling, Dr. Alfried Kiehl
II-3	Test methods	Peter Wißling, Dr. Alfried Kiehl
III	Special effect pigments	
III-1	PVD pigments	Dr. Jörg Seubert
III-2	Coloured aluminium pigments	Kerstin Ahne, Hermann Birner, Dr. Frank Henglein
IV	Applications	
IV-1	Coating systems and their applications	
IV-1.1	Coating systems	
IV-1.1.1	Solvent-based coatings	Peter Wißling
IV-1.1.2	Water-based coatings	Alexander Albrecht, Dr. Alfried Kiehl, Peter Wißling
IV-1.1.3	Powder coatings	Burkhard Schreiber
IV-1.1.4	UV coatings	Peter Wißling
IV-1.2	Applications	
IV-1.2.1	Automotive coatings	Dr. Andrea Fetz, Peter Wißling
IV-1.2.2	Decorative coatings	Robert Maul
IV-1.2.3	Industrial coatings and functional applications	Dr. Ulrich-Andreas Hirth
IV-2	Print	
IV-2.1	Offset	Dr. Mark Stoll
IV-2.2	Gravure, flexographic and screen printing	Dr. Jörg Seubert
IV-3	Plastics	Martin Schaal
IV-4	Technical applications	Jürgen Taraba, Uwe Hartmann
IV-5	Corrosion protection	Dr. Andrea Fetz, Peter Wißling
IV-6	Cosmetics and personal care products	Martha Graziano, Dr. Günter Kaupp, Dr. Peter Krüger
V	Effects on people and environment	Dr. Angelika Hartl-Gunselmann, Thomas Hartmann
VI	Safety advice	Dr. Dieter Lutz

Kerstin Ahne studied Industrial Chemistry at the Georg-Simon-Ohm Technical College in Nuremberg. Since 1999 she has been Project Manager at Eckart and is since 2003 Group Leader in the Applications Technology Department for Powder Coatings and Bonding.

Alexander Albrecht studied Industrial Chemistry at the Georg-Simon-Ohm Technical College in Nuremberg. Since 1998, he is employed at Eckart as Project Manager in Research & Development.

Hermann Birner has been trained as a Chemical Laboratory Assistant and subsequently as a Chemical Technician. He is employed at Eckart since 1994, first in Research & Development and currently as Project Manager for a pilot plant line.

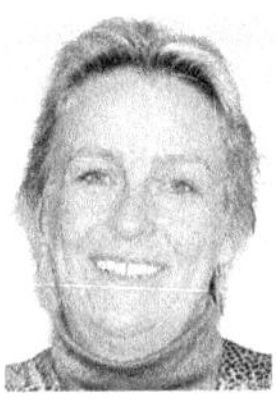

Dr. Andrea Fetz studied Chemistry at Friedrich-Alexander University Erlangen-Nuremberg and subsequently obtained a doctorate. She is working at Eckart since 1988, initially in Research & Development and is now being employed in Marketing Coatings.

Martha Graziano studied biology in the USA. From1988 to 2001 she was employed at Süd Chemie Rheologicals (USA). Since 2001 she works at Eckart America in the Personal Care Department.

Dr. Angelika Hartl-Gunselmann read Chemistry at the Friedrich-Alexander University Erlangen-Nuremberg. Employed by Eckart since 1990, she is heading the Legal Group for Products and Patents since 1993.

Thomas Hartmann graduated as a Chemical Engineer from Friedrich-Alexander-University Erlangen-Nuremberg. He is working in Eckart as Safety Engineer since 1989.

After training as a Pharmaceuticals Technician, **Uwe Hartmann** studied Industrial Chemistry at University for Applied Science in Nuremberg. He is being employed at Eckart since 1994 and is working in the Aluminium Flakes Application Technology Department since 1998.

Dr. Frank Henglein read Chemistry at the Technical University in Berlin and at Bayreuth University. From 1994 to 1997 he was employed at the Research Institute Juelich; since 1997 he is working at Eckart in R & D.

Ulrich-Andreas Hirth studied chemistry at the University of Wuerzburg. After his PHD in 1991 he started working at Eckart and was responsible for Technical Service for metallic pigments for the printing industry. Now he is responsible for the marketing of effect pigments for industrial and powder coating applications.

Dr. Günter Kaupp studied Chemistry at the Julius-Maximilians-University in Wuerzburg. He joined Eckart in 1993 and is responsible for development of nacreous pigments and technical coordination since 1998.

Dr. Alfried Kiehl studied Chemistry at the University of Mainz. This was followed by two years at the Politecnico di Milano (Italy) in polymer chemistry. He has been at Eckart since 1995, initially as a Project Leader in R & D, and is currently working in Marketing.

Dr. Peter Krüger read Chemistry at Julius-Maximilians-University in Wuerzburg. Since 1998 he is working at Eckart in Research & Development Pearlescent Pigments. Since 2004 he is also responsible for Cosmetic Applications.

Dr. Dieter Lutz graduated as a Chemical Engineer from University of Erlangen-Nuremberg. He joined Eckart in 1997 initially as a process engineer. After further qualifications in Safety Engineering, he became head of the Department for Health and Safety in 2001.

After training as a Laboratory Technician, **Robert Maul** studied at Technical College Niederrhein in Krefeld. He was Laboratory Manager from 1992 until 1995 at Reicolor. Since 1995 he has been employed at Eckart and is at present responsible for Marketing of Industrial Coatings.

Martin Schaal studied Chemistry at the Technical College Aalen. From 1980 to 1985 he worked at BASF and from 1985 to 1994 at Great Lakes Chemicals. Since 1994, he is employed at Eckart in the Applications Technology Department for Effect Pigments in Plastics.

Burkhard Schreiber studied in Paderborn. From 1988 to 1996 he was Laboratory Manager at Ruhrpulver, From 1996 to 1999 he worked as a salesman for various pigment manufacturers. Since 1999 he is employed at Eckart in the Applications Technology Department for Powder Coatings and Bonding.

Dr. Jörg Seubert read Chemistry at the Friedrich-Alexander University Erlangen-Nuremberg and has worked at Eckart since1995. He is at present responsible for Product Management of metallic effect pigments and Marketing of Graphic Arts products.

Dr. Mark Stoll read Chemistry at the Friedrich-Alexander University Erlangen-Nuremberg. He has been at Eckart since 1996 and is responsible for Marketing in the Plastic Coatings Unit since 2002.

Jürgen Taraba studied Chemistry at the Engineering College Magdeburg. From 1966 to 1967 he worked at Wolfen and from 1968 to 1991 at Leichtmetallwerk Rackwitz. He has been at Eckart since 1992 and is at present responsible for Marketing of Technical Aluminiumflakes.

Peter Wissling managing author of this book, studied Chemistry and Technology of Coating Materials in Paderborn. He subsequently worked for Heubach and Morton. Since 1992 he has been employed at Eckart in the Technical Services Unit for Liquid Paints. Presently he looks after Northern Europe and Asia Pacific region.

Acknowledgement

As managing author, who has followed this book from the beginning on through all stages, it is now my part to say "thank you". Many people, not all of whom can be individually mentioned here, contributed to the success of the work and special thanks are due to them.

Particular thanks go to the management of Eckart GmbH & Co. KG who gave permission for this project without hesitation. Dr Klaus Greiwe, author of the foreword, and Dr Ostertag contributed with helpful and good discussions and suggestions. Because of the huge range of topics, it was not possible to entrust the writing of this book to a single author and so a further twenty people contributed to it. The fact that this was done mainly outside the normal course of their duties is gratefully recognised and acknowledged. In particular, thanks are due to Simone Grallath, Sabine Kunzog and Melanie Quenzler who undertook the detailed editorial work in a critical and responsible manner. I would also like to thank them for many, often extensive, discussions.

I should also mention the translators whose rendering of technical German into English has been done excellently. This book would never have appeared without the initiative of Dr Ulrich Zorll. With experience, patience and many suggestions he was the driving force behind this project. The whole team owes him a lot.

Velden/Germany, April 2006

Peter Wißling

Index

G

H

I

J

L

M

N

O

P

R

S

T

W

www.ingramcontent.com/pod-product-compliance
Lightning Source LLC
LaVergne TN
LVHW021727021026
849161LV00004BA/759
9783878701712